Pro|ENGINEER®
WILDFIRE™ 2.0

MECHANICA Tutorial
(Structure / Thermal)

Integrated Mode

Roger Toogood, Ph.D., P. Eng.
Edmonton, Alberta

ISBN: 1-58503-190-9

SDC
PUBLICATIONS

Schroff Development Corporation

www.schroff.com
www.schroff-europe.com

DISCLAIMER

The discussion, examples, and exercises in this tutorial are meant only to demonstrate the functionality of the program and are not to be construed as fully engineered design solutions for any particular problem. Use of the methods and procedures described herein are for instructional purposes only and are not warranted or guaranteed to provide satisfactory solutions in any specific application. The author and publisher assume no responsibility or liability for any results or solutions obtained using the methods and procedures described herein.

MECHANICA® Tutorial
(Structure / Thermal)
Integrated Mode

Updated For

Pro | ENGINEER®
W I L D F I R E™ 2.0

Preface

In his excellent text <u>Finite Element Procedures,</u> K.J. Bathe identifies two possible and different objectives for studying Finite Element Analysis (FEA) and methods: to learn the proper use of the method for solving complex problems (the practitioner's goal), and to understand the methods themselves in depth so as to pursue further development of the theory (the researcher's goal). This tutorial was created with the former objective in mind, recognizing that this is a formidable task and not one that can be totally accomplished in a single, short volume. Thus, the primary purpose of the tutorial is to introduce new users to MECHANICA® (Parametric Technology Corporation, Waltham, MA) and see how it can be used to analyze a variety of problems.

The tutorial lessons cover most of the major concepts and frequently used commands required to progress from a novice to an intermediate user level. The commands are presented in a click-by-click manner using simple examples and exercises that illustrate a broad range of the most common analysis types that can be performed. In addition to showing/illustrating the command usage, the text will explain why certain commands are being used and, where appropriate, the relation of commands to the overall FEA philosophy. Moreover, since error analysis is an important skill, considerable time is spent exploring the created models (in fact, sometimes intentionally inducing some errors), so that users will become comfortable with the "debugging" phase of modeling.

In this 7th edition, the tutorial has been updated to Wildfire 2.0, the latest release of Pro/ENGINEER (March, 2004). The tutorial deals exclusively with operation in integrated mode with Pro/ENGINEER. A companion book from SDC deals with using MECHANICA in independent mode. Many problem types (in addition to solids) can be treated in integrated mode. These include 2D models (plane stress/strain and axisymmetric solids and shells) created using the Pro/E interface. Shell and beam idealizations are also possible. Other recent enhancements to the program include cyclic constraints, changes in the interface and operation of the program such as the inclusion of simulation features in the model tree. These are all covered in this tutorial. The capability (introduced in Release 2000i) to handle large deformation problems, that is problems involving geometric non-linearity, has not been covered due to space constraints. In any case, this functionality is of a significantly more advanced level than what is required (perhaps) in an introductory tutorial.

The Tutorial has been carefully edited to reflect the significant changes that have occurred in the Wildfire 2.0 interface. Many aspects of that interface are covered in the Pro/ENGINEER Wildfire 2.0 Tutorial (also from SDC) and are not repeated here (for example, location and operation of the model tree in the Navigator window). Because this book deals exclusively with MECHANICA operating in integrated mode, familiarity with Pro/ENGINEER is assumed. Although references are sometimes made to independent mode operation, this is not treated here.

Students with a broad range of backgrounds should be able to use this book. The lessons were written for new users with no previous experience with FEA, although some familiarity with computers and elementary strength of materials is necessary. Because the emphasis is on MECHANICA, the Pro/E models are not too complex and should be easily created by novice users of that program. To save some time, geometric models may be downloaded from the SDC web site at <**www.schroff1.com**>.

This book is **NOT** a complete reference manual for MECHANICA. There are several thousand pages of reference manuals available on-line with the MECHANICA installation, with good search tools and cross-referencing to allow users to find relevant material quickly. There have been significant improvements in the Help system with Wildfire to make this task easier.

The tutorial treats solid models first, as these are the default model type created in Pro/E. This is followed later by model idealizations: plane stress, plane strain, shells, beams, frames, and axisymmetric shells and solids, springs, masses, and so on. A new chapter has been added that introduces the tools for solving simple steady and transient thermal problems, and for creating temperature loads used to compute thermally induced stresses.

It continues to be a challenge to decide what to include and what to exclude in this introduction in terms of the command set within MECHANICA. The author can only hope that the presented material will be found useful, and in the right dose! It has also been interesting to design suitable demonstration problems and exercises that are interesting, feasible with the state of learning of the user, physically meaningful, and illustrate a broad set of MECHANICA functionality - all within the space of 200 or so pages. It is hoped that at least some of these goals have been satisfied.

Although every effort has been made in proofreading the text, it is inevitable that errors will appear. The author takes full responsibility for these and hopes they will not impede your progress through the tutorial. Any comments, criticisms, and/or suggestions will be gratefully received and acknowledged. You can reach the author by email at <*procaden@telusplanet.net*>.

Enjoy the book!

Notes to Instructors:

The tutorials consist of the following:

> 2 lessons on general introductory material (reading only, but important!)
> 2 lessons introducing the basic operations in Pro/M using solid models
> 4 lessons on model idealizations (shells, beams and frames, plane stress, etc)
> 1 lesson on miscellaneous topics
> 1 lesson on steady and transient thermal analysis

Each of these tutorial chapters will take between 1-1/2 to 3 hours to complete depending on the ability and background of the student. Moreover, additional time would be beneficial for experimentation and additional exploration of the program. Most of the material can be done by the students on their own, however there are a few "tricky" bits in some of the lessons. Therefore, it is important to have experienced and knowledgeable teaching assistants available (preferably right in the computer lab) who can answer special questions and especially bail out students who get into trouble. Most common causes of confusion are due to not completing the lessons or digesting the material. This is not surprising given the volume of new information or the lack of time in students' schedules. However, I have found that most student questions are answered within the lessons.

In addition to the tutorials, it is presumed that some class time over the duration of a course will be used for discussion of some of the broader issues of FEA, such as the treatment of constraints. Furthermore, it is vitally important for students to compare their FEA results with other possible solutions. This can be accomplished using simple problems for which either analytical solutions or experimental data exist. An extended discussion and exploration of modeling of boundary conditions would be very beneficial, particularly in cases (I call them "diabolical") where seemingly reasonable modeling alternatives can produce significantly different results. It takes a while for students to realize that just creating the model and producing pretty pictures is not sufficient for design work, and the notions of accuracy and convergence need careful treatment and discussion. On the other hand, examining case after case of diabolical models that produce dubious answers can have the unintended effect of turning students off the method entirely.

It should be expected that most students, after having gone through a lesson only once, will not have "internalized" very much. My experience is that many students execute the commands without carefully reading or studying the accompanying text explaining *why*. The second pass through the lesson usually results in considerably more retention and understanding. Each lesson concludes with a number of review questions and simple (?) exercises that can be completed using commands taught in that lesson. Where possible, students should be given additional problems that can be verified independently by experiment or analytical methods. Students really don't feel comfortable or confident until they can make models from scratch on their own.

That having been said, I am continually amazed at how quickly many students can get up the learning curve on both Pro/ENGINEER and MECHANICA. Any instructor introducing this software to a group of capable and interested students should be prepared to move very quickly to stay ahead of the class!

Acknowledgments

Some of the models used in these tutorials are based on the treatment in **The Finite Element Method in Mechanical Design** (PWS-Kent, 1993) by Charles E. Knight. This is a clearly written and informative book, although emphasis is on the h-code analysis with only tangential mention of the p-code method used in MECHANICA.

Thanks are due to Stephen Schroff and Mary Schmidt at Schroff Development Corporation and Janet Drumm at JourneyEd for their efforts in taking this work to a wider audience and for their tolerance of the delays in its arrival!

Once again I must express special thanks to Elaine, Jenny, and Kate for their patience while I was occupied with this work.

To users of this material, I hope you enjoy the lessons. I apologize beforehand for any omissions and errors that may have appeared and I would appreciate any comments, criticisms, and suggestions for the improvement of this manual.

RWT
Edmonton, Alberta
4 July 2004

Organization and Synopsis of the Tutorials

A brief synopsis of the ten chapters in this book is given below. Each chapter should take at least 1.5 to 3 hours to complete - if you go through the lessons too quickly or thoughtlessly, you may not understand or remember the material. For best results, it is suggested that you scan/browse through the lesson or major section completely before going through it in detail. You will then have a sense of where the lesson is going, and not be tempted to just follow the commands blindly. You need to have a sense of the forest when examining each individual tree!

Chapter 1 - Introduction to MECHANICA

An introduction to finite element analysis, with some cautions about its use and misuse; examples of problems solved with MECHANICA; organization of the tutorials; tips and tricks for using MECHANICA

Chapter 2 - Finite Element Modeling with MECHANICA

Background information on FEA. The concept of modeling. Particular attention is directed at concerns of accuracy and convergence of solutions, and the differences between h-code and p-code FEA. Overview of MECHANICA operations and nomenclature

Chapter 3 - Solid Models (Part 1 - Static Analysis)

A simple model is created using Pro/ENGINEER and analyzed in MECHANICA. The complete sequence of steps required for a static analysis will be outlined, and basic result display options presented. Automatic mesh generation.

Chapter 4 - Solid Models (Part 2 - Sensitivity Studies and Optimization)

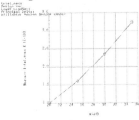

Design parameters are designated for sensitivity studies and optimization. Special concerns for applying loads and constraints on solid models are explored. Superposition and multiple load sets are introduced.

Chapter 5 - Plane Stress and Plane Strain

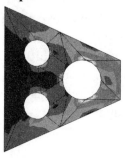

This is the first lesson on idealizations. It deals with problems that can be classed as either plane stress or plane strain. In either case, the model is based on 2D geometry that can be extracted from the Pro/E solid model. The use of symmetry is introduced.

Chapter 6 - Axisymmetric Solids and Shells

Axisymmetric models are another case where a 2D idealization can be used. Two are available: axisymmetric solids and shells. New load types are introduced: centrifugal and thermal loads

Chapter 7 - Shell Models

Shell models are an idealization for general 3D models that can be used when a part is composed of thin-walled features. Shell geometry can be created automatically or manually. Shells can also be combined with solids in the same model. Bearing loads are introduced. Convergence problems are discussed.

Chapter 8 - Beams and Frames

Beam elements are the final idealization. Using beams requires a good understanding of beam coordinate systems, sections, and orientation. Point and distributed loads are covered, as are beam releases. Preparation of shear and bending moment diagrams.

Chapter 9 - Miscellaneous Topics

Several topics are introduced here, starting with cyclic symmetry. The use of springs and masses is examined. Modal analysis is introduced. Finally, the use of contact surfaces in a simple assembly is demonstrated.

Chapter 10 - Thermal Analysis

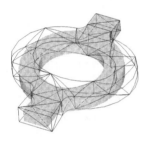

An introduction to steady state and transient thermal analysis. The importance of units. Temperature distribution in a solid and 2D plate model. Idealizations. Transient thermal problems. Using Thermal results in Structure.

TABLE OF CONTENTS

Chapter 1 - Introduction to the Tutorials

Chapter 2 - Finite Element Modeling with MECHANICA

Chapter 3 - Solid Models (Part 1)

Chapter 4 - Solid Models (Part 2)

Chapter 5 - Plane Stress and Plane Strain Models

Chapter 6 - Axisymmetric Solids and Shells

Chapter 7 - Shell Models

Chapter 8 - Beams and Frames

Chapter 9 - Miscellaneous Topics

Chapter 10 - Thermal Models

Chapter 1 :

Introduction to the Tutorials

Synopsis

An introduction to finite element analysis, with some cautions about its use and misuse; examples of problems; organization of the tutorials; tips and tricks for using MECHANICA

Overview of this Lesson

This lesson will be used to get you set up for the rest of the tutorials - to set the stage for what is to come. It will go over some basic ideas about FEA and what you can do with MECHANICA. The lesson is quite short, and will cover the following:

- ♦ general comments about using Finite Element Analysis (FEA)
- ♦ examples of problems solved using MECHANICA Structure
- ♦ layout of the tutorials
- ♦ how the tutorial will present command sequences
- ♦ some tips and tricks for using MECHANICA

Finite Element Analysis

Finite Element Analysis (FEA), also known as the Finite Element Method (FEM), is probably the most important tool added to the mechanical design engineer's toolkit in recent years. The development of FEA has been driven by the desire for more accurate design computations in more complex situations, allowing improvements in both the design procedure and products. The growing use of FEA has been made possible by the creation of affordable computers that are capable of handling the immense volume of calculations necessary to prepare and carry out an analysis and easily display the results for interpretation. With the advent of very powerful desktop workstations, FEA is now available at a practical cost to virtually all engineers and designers.

The MECHANICA software described in this introductory tutorial is only one of many commercial systems that are available. All of these systems share many common capabilities. In this tutorial, we will try to present both the commands for using MECHANICA and the reasons behind those commands, so that the general procedures can be transferred to other FEA packages. Notwithstanding this desire, it should be realized that MECHANICA is unique in many ways among packages currently available. Therefore, numerous topics treated will be specific to MECHANICA.

MECHANICA is actually a suite of three programs: *Structure*, *Thermal*, and *Motion*. The first of these, *Structure*, is able to perform the following:

- ◆ linear static stress analysis
- ◆ modal analysis (mode shapes and natural frequencies)
- ◆ buckling analysis
- ◆ large deformation analysis (non-linear)

and others. This manual will be concerned only with the first two of these analyses. The remaining types of problems are beyond the scope of an introductory manual. Once having finished this manual, however, interested users should not find the other topics too difficult. The other two programs (*Thermal* and *Motion*) are used for thermal analysis and dynamic analysis of mechanical systems, respectively. Both of these programs can pass information (for example temperature distributions) back to *Structure* in order to compute the associated stresses. This book contains an introduction to *Thermal* for analysis of simple steady and transient problems in heat transfer. The final program, *Motion*, is not treated here. Some of the functionality of that program has been included within Wildfire itself (mechanism kinematics, with limited dynamic capability) with the appropriate license configuration (MDX -Mechanism Design Extension). Full dynamic simulation capabilities remain the domain of *Motion*.

MECHANICA offers much more than simply an FEA engine. We will see that MECHANICA is really a design tool since it will allow parametric studies as well as design optimization to be set up quite easily. Moreover, unlike many other commercial FEM programs where determining accuracy can be difficult or time consuming, MECHANICA will be able to compute results with some certainty as to the accuracy[1].

MECHANICA does not currently have the ability to handle non-linear problems, for example a stress analysis problem involving a non-linearly elastic material like rubber. However, as of Release 2000*i*, problems involving very large geometric deflections can be treated, as long as the stresses remain within the linearly elastic range for the material.

In this tutorial, we will concentrate on the main concepts and procedures for using the software and focus on topics that seem to be most useful for new users and/or students doing design projects and other course work. We assume that readers do not know anything about the software, but are quite comfortable with Pro/ENGINEER. A short and very qualitative overview of the FEA theoretical background has been included, but it should be emphasized that this is very limited in scope. Our attention here is on the use and capabilities of the software, not providing a complete course on using FEA, its theoretical origins, or the "art" of FEA modeling strategies. For further study of these subjects, see the reference list at the end of the second chapter.

---

[1]This refers to the problem of "convergence" whereby the FEA results must be verified or tested so that they can be trusted. We will discuss convergence at some length later on and refer to it continually throughout the manual.

Examples of Problems Solved using MECHANICA

To give you a taste of what is to come, here are three examples of what you will be able to do with MECHANICA on completion of these tutorials. The examples, all solved with *Structure*, are a simple analysis, a parametric design study called a sensitivity analysis, and a design optimization. In MECHANICA's language, these are called *design studies*.

Example #1 : Analysis

This is the "bread and butter" type of problem for MECHANICA. A model is defined by some geometry (in 2D or 3D) in the geometry pre-processor. This is not as simple or transparent as it sounds, as discussed below. The model is transferred into MECHANICA where material properties are specified, loads and constraints are applied, and one of several different types of analysis can be run on the model. In the figure at the right, a model of a somewhat crude connecting rod is shown. This part is modeled using 3D solid elements. The hole at the large end is fixed and a lateral bearing load is applied to the inside surface of the hole at the other end. The primary results are shown in Figures 2 and 3. These are

Figure 1 Solid model of a part

contours of the Von Mises stress[2] on the part, shown in a *fringe* plot (these are, of course, in color on the computer screen), and a wireframe view of the total (exaggerated) deformation of the part (this can be shown as an animation). Here, we are usually interested in the value and location of the maximum Von Mises stress in the part, whether the solution agrees with our desired boundary conditions, and the magnitude and direction of deformation of the part.

Figure 2 Von Mises stress fringe plot

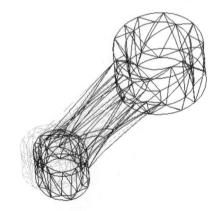

Figure 3 Deformation of the part

---

[2] The Von Mises stress is obtained by combining all the stress components at a point in a way which produces a single value that can be compared to the yield strength of the material. This is the most common way of examining the computed stress in a part.

Example #2 : Sensitivity Study

Often you need to find out the overall effect on the solution of varying one or more design parameters, such as dimensions. You could do this by performing a number of similar analyses, and changing the geometry of the model between each analysis. MECHANICA has an automated routine which allows you to specify the parameter to be varied, and the overall range. It then automatically performs all the modifications to the model, and computes results for the intermediate values of the design parameters.

The example shown in Figure 4 is a quarter-model (to take advantage of symmetry) of a transition between two thin-walled cylinders. The transition is modeled using shell elements.

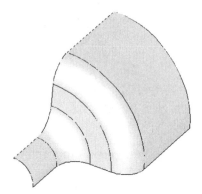

Figure 4 3D Shell quarter-model
of transition between cylinders

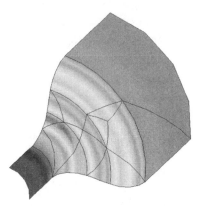

Figure 5 Von Mises stress in
shell model

Figure 5 shows the contours of the Von Mises stress on the part. The maximum stress occurs at the edge of the fillet on the smaller cylinder just where it meets the intermediate flat portion. The design parameter to be varied is the radius of this fillet, between the minimum and maximum shapes shown in Figures 6 and 7.

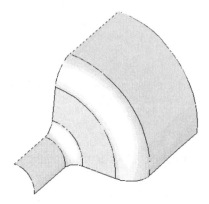

Figure 6 Minimum radius fillet

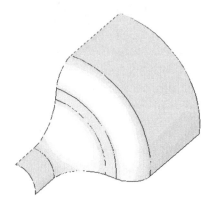

Figure 7 Maximum radius fillet

Figure 8 shows the variation in the maximum Von Mises stress in the model as a function of radius of the fillet. Other information about the model, such as total mass, or maximum deflection is also readily available, also as a function of the radius.

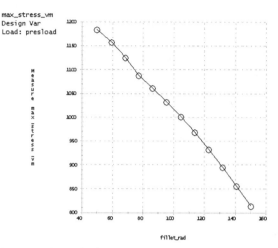

Figure 8 Variation of Von Mises stress with fillet radius in shell model

Example #3 : Design Optimization

This capability of MECHANICA is really astounding! When a model is created, some of the geometric parameters can be designated as design variables. Then MECHANICA is turned loose to find the combination of values of these design variables that will minimize some objective function (like the total mass of the model) subject to some design constraints (like the allowed maximum stress and/or deflection). MECHANICA searches through the design space (for specified ranges of the design variables) and will find the optimum set of design variables automatically!

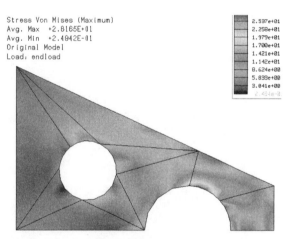

Figure 9 Initial Von Mises stress distribution in plate before optimization

The example shown is of a plane stress model of a thin, symmetrical, tapered plate under tension. The plate is fixed at the left edge, while the lower edge is along the plane of symmetry. A uniform tensile load is applied to the vertical edge on the right end. The Von Mises stress contours for the initial design are shown in Figure 9. The maximum stress, which exceeds a design tolerance, has occurred at the large hole located on the plate centerline, at about the 12:30 position. The stress level around the smaller hole is considerably less, and we could probably increase the diameter of this hole in order to reduce mass. The question is: how much?

The selected design variables are the radii of the two holes. Minimum and maximum values for these variables are indicated in the Figures 10 and 11. The objective of the optimization is to minimize the total mass of the plate, while not exceeding a specified maximum stress.

Figure 10 Minimum values of design variables

Figure 11 Maximum values of design variables

Figure 12 shows a history of the design optimization computations. The figure on the left shows the maximum Von Mises stress in the part that initially exceeds the allowed maximum stress, but MECHANICA very quickly adjusts the geometry to produce a design within the allowed stress. The figure on the right shows the mass of the part. As the optimization proceeds, this is slowly reduced until a minimum value is obtained (approximately 20% less than the original). MECHANICA allows you to view the shape change occurring at each iteration.

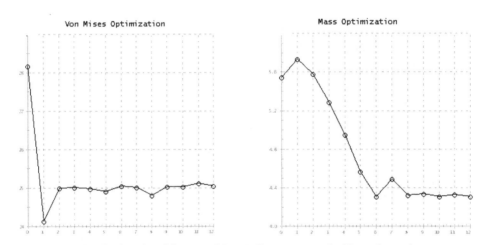

Figure 12 Optimization history: Von Mises stress (left) and total mass (right)

The final optimized design is shown in Figure 13. Notice the increased size of the interior hole, and the more efficient use of material. The design limit stress now occurs on both holes.

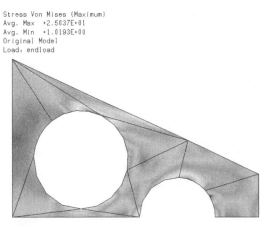

Stress Von Mises (Maximum)
Avg. Max +2.5037E+01
Avg. Min +1.0193E+00
Original Model
Load: endload

Figure 13 Von Mises stress distribution in
optimized plate

In these lessons, we will explore variations of these three types of design study (simple analysis, sensitivity studies, and optimization). We will explore a number of different types of models while doing this (solids, shells, beams, plates, etc.).

FEA User Beware!

Users of this (or any other FEA) software should be cautioned that, as in other areas of computer applications, the GIGO ("Garbage In = Garbage Out") principle applies. Users can easily be misled into blind acceptance of the answers produced by the programs. **Do not confuse pretty graphs and pictures with correct modeling practice and accurate results**.

A skilled practitioner of FEA must have a considerable amount of knowledge and experience. The current state of sophistication of CAD and FEA software may lead non-wary users to dangerous and/or disastrous conclusions. Users might take note of the fine print that accompanies all FEA software licenses, which usually contains some text along these lines: "The supplier of the software will take no responsibility for the results obtained . . ." and so on. Clearly, the onus is on the user to bear the burden of responsibility for any conclusions that might be reached from the FEA.

We might plot the situation something like Figure 14 on the next page. In order to intelligently (and safely) use FEA, it is necessary to acquire some knowledge of the theory behind the method, some facility with the available software, and a great deal of modeling experience. In this manual, we assume that the reader's level of knowledge and experience with FEA initially places them at the origin of the figure. The tutorial (particularly Chapter 2) will extend your knowledge a little bit in the "theory" direction, at least so that we can know what the software requires for input data, and (generally) how it computes the results. The step-by-step tutorials and exercises will extend your knowledge in the "experience" direction. Primarily, however, this tutorial is meant to extend your knowledge in the "FEA software" direction, as it applies to using MECHANICA. Readers who have already moved out along the "theory" or "experience" axes will have to bear with us - at least this manual should help you discover the capabilities of the MECHANICA software package.

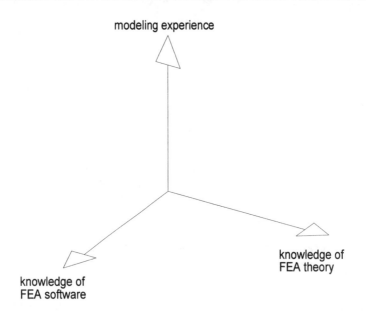

Figure 14 Knowledge, skill, and experience requirements for FEA users

In summary, some quotes from speakers at an FEA panel at an ASME Computers in Engineering conference in the early 1990's should be kept in mind:

"Don't confuse convenience with intelligence."
In other words, as more powerful functions get built in to FEA packages (such as automatic or even adaptive mesh generation), do not assume that these will be suitable for every modeling situation, or that they will always produce trustworthy results. If an option has defaults, be aware of what they are and their significance to the model and the results obtained. Above all, remember that just because it is easy, it is not necessarily right!

"Don't confuse speed with accuracy."
Computers are getting faster and faster. This also means that they can compute an inaccurate model faster than before - a wrong answer in half the time is hardly an improvement!

and finally, the most important:

"FEA makes a good engineer better and a poor engineer dangerous."
As our engineering tools get more sophisticated, there is a tendency to rely on them more and more, sometimes to dangerous extremes. Relying solely on FEA for design verification might be dangerous. Don't forget your intuition, and remember that a lot of very significant engineering design work has occurred over the years on the back of an envelope. Let FEA become a tool that extends your design capability, not define it.

Layout of this Manual

Running the MECHANICA software is not a trivial operation. However, with a little practice, and learning only a fraction of the capabilities of the program, you can perform FEA of reasonably complex problems. This manual is meant to guide you through the major features of the software and how to use it. It is not meant to be a complete guide to either the software or FEA modeling - consider it the elementary school of practical FEA!

Chapter 2 of the tutorial will present an overview of the theory and mathematics behind how FEA is implemented in MECHANICA. In particular, the origin and differences between h-code analysis and the p-code method in MECHANICA are discussed. The primary purpose of this chapter is to outline the main capabilities of MECHANICA as they apply to the design and analysis of mechanical parts. These include simple analyses, sensitivity studies, and parameter optimization. This chapter will basically introduce you to the terminology used in the program, and give you an overview of its operation.

Chapters 3 and 4 will present the basic procedure and commands for performing design studies on solid models. This is a natural starting point, given that models imported from Pro/E are usually solids. Common methods of displaying results are shown. Some issues of modeling are discussed, such as symmetry. Several modeling pitfalls, which also occur in other model types are investigated, and solutions proposed.

Chapter 5 will introduce you to the analysis of 2D models using idealizations. These are plane stress and plane strain analyses. Geometry for these models is selected from the 3D part geometry as created in Pro/E. The idealization, when applicable, results in a significant reduction in the computational effort for the model.

The subject of Chapter 6 is axisymmetric models. These require that the geometry, loads, and constraints can be based on a 2D layout that represents the problem.

Chapter 7 is devoted to a very important idealization - the shell model. Shells occur when the model contains all or some thin-walled solid features. This idealization results in a greatly reduced problem size and faster solution.

Beams and 2D and 3D frames (including trusses) are dealt with in Chapter 8. Both single continuous beams and beams as components of frames are discussed. Beams can also be used in combination with shells and solids.

Chapter 9 will deal with some miscellaneous topics in *Structure* including cyclic symmetry, spring and mass elements, modal analysis, and contact analysis in assemblies.

The final chapter will introduce methods in *Thermal* to solve heat transfer problems in 2D and 3D geometries. These can be steady state or transient solutions that allow computation of temperature distributions, heat flux through the model, or heat transfer from the boundary. You must be especially careful with units in this module. Temperature distributions can be taken back into *Structure* to compute thermally induced stresses.

At the end of each of these chapters, a number of additional exercises are presented. You should try to do as many of these as you can in order to build up your knowledge and repertoire of modeling scenarios.

Tips for using MECHANICA

In the tutorial examples that follow, you will be lead through a number of simple problems keystroke by keystroke. Each command will be explained in depth so that you will know the "why" as well as the "what" and "how". Resist the temptation to just follow the keystrokes - you must think hard about what is going on in order to learn it. You should go through the tutorials while working on a computer so that you experience the results of each command as it is entered. Not much information will sink in if you just read the material. We have tried to capture exactly the key-stroke, menu selection, or mouse click sequences to perform each analysis. These actions are indicated in **_bold face italic type_**. Characters entered from the keyboard are enclosed within square brackets. When more than one command is given in a sequence, they are separated by the symbol ">". When several commands are entered on a single menu or window, they are separated by the pipe symbol " | ". An option from a pull-down list will be indicated with the list title and selected option in parantheses. So, for example, you might see command sequences similar to the following:

> **_Materials > Assign > Part > STEEL_IPS | Accept_**
> **_Analysis (QuickCheck)_**
> **_Results > Create > [VonMises] | Accept_**

At the end of each chapter in the manual, we have included some Questions for Review and some simple Exercises which you should do. These have been designed to illustrate additional capabilities of the software, some simple modeling concepts, and sometimes allow a comparison with either analytical solutions or with alternative modeling methods. The more of these exercises you do, the more confident you can be in setting up and solving your own problems.

Finally, here a few hints about using the software. Menu items and/or graphics entities on the screen are selected by clicking on them with the _left mouse button_. We will often refer to this as a 'left click' or simply as a 'click'. The _middle mouse button_ ('middle click') can be used (generally) whenever **_Accept, Enter, Close_** or **_Done_** is required. The dynamic view controls are obtained using the mouse as shown in Table 1. Users of Pro/E will be quite comfortable with these mouse controls. Any menu commands grayed out are unavailable for the current context. Otherwise, any menu item is available for use. You can, for example, jump from the design menus to the pulldown menus at any time. Many operations can be launched by clicking and holding down the right mouse button on an entry in the model tree or in the graphics window. This will produce a (context sensitive) pop-up menu of relevant commands.

Table 1-1 Common Mouse Functions

Function		Operation	Action
Selection (click left button)		LMB	entity or command under cursor selected
Direct View Control (drag holding middle button down)		MMB	Spin
		Shift + MMB	Pan
		Ctrl + MMB (drag vertical)	Zoom
		Ctrl + MMB (drag horizontal)	Rotate around axis perpendicular to screen
		Roll MMB scroll wheel (if available)	Zoom
Pop-up Menus (click right button)		RMB with cursor over blank graphics window	launch context-sensitive pop-up menus

As of Release 2001, Pro/E and MECHANICA incorporated a new "object-action" operating paradigm (as opposed to the previous "action-object" form). This means you can pick an object on the screen (like a part surface), then specify the action to be performed on it (like applying a load). This is a much more streamlined and natural sequence to process commands. Of course, the previous action-object form will still work. In this Tutorial, command sequences are represented at various times in either of the two forms. Hopefully, this will not get confusing.

So, with all that out of the way, let's get started. The next chapter will give you an overview of FEA theory, and how MECHANICA is different from other commercial packages.

Questions for Review

1. In MECHANICA-ese, what is meant by a "design study?"
2. What are the three types of design study that can be performed by MECHANICA?
3. What is the Von Mises stress? From a strength of materials textbook, find out how this is computed and its relation to yield strength. Also, for what types of materials is this a useful computation?
4. Can MECHANICA treat non-linear problems?
5. What does GIGO mean?
6. What three areas of expertise are required to be a skilled FEA practitioner?

Exercises

1. Find some examples of cases where seemingly minor and insignificant computer-related errors have resulted in disastrous consequences.

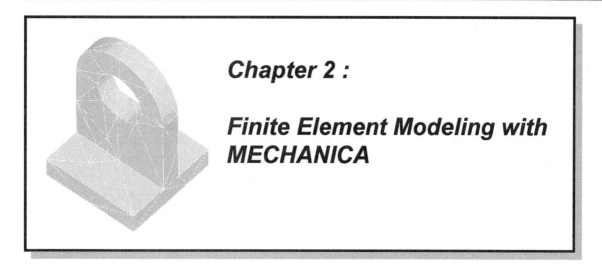

Chapter 2 :

Finite Element Modeling with MECHANICA

Synopsis

Background information on FEA. The concept of modeling. Accuracy and convergence of solutions. Differences between h-code and p-code FEA. Overview of MECHANICA.

Overview of this Lesson

This chapter presents an overall view of FEA in general, and discusses a number of ideas and issues involved. The major differences between Pro/M, which uses a p-code method, and other packages, which typically use h-code, are presented. The topics of accuracy and convergence are discussed. The major sections in this chapter are:

- overview and origins of FEA
- discussion of the concept of the "model"
- general procedure for FEA solutions
- FEA models versus CAD models
- p-elements and h-elements
- convergence and accuracy
- sources of error
- overview of MECHANICA

Although you are probably anxious to get started with the software, your understanding of the material presented here is very important. We will get to the program soon enough!

Finite Element Analysis : An Introduction

In this section, we will try to present the essence of FEA without going into a lot of mathematical detail. This is primarily to set up the discussion of the important issues of accuracy and convergence later in the chapter. Some of the statements made here are generalizations and oversimplifications, but we hope that this will not be too misleading. Interested users can consult any of a number of text and reference books (some are listed at the end of this chapter) which describe the theoretical underpinnings of FEA in considerably greater detail.

In the following, the ideas are illustrated using a planar (2D) solution region, but of course these ideas extend also to 3D. We'll try to keep the mathematics to a minimum here! Let's suppose that we are faced with the following problem: We are given a body or connected region (or volume) R with a boundary B as shown in Figure 1(a). Some continuous physical variable, e.g. temperature T or displacement u, is governed by a physical law within the body and subjected to known conditions on the boundary. This physical law might come from heat conduction or the equations of elasticity. In a finite element solution, the geometry of the region is typically generated by a CAD program, such as Pro/ENGINEER.

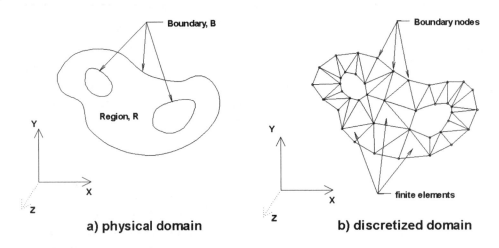

Figure 1 The problem to be solved is specified in a) the physical domain
and b) the discretized domain used by FEA

For a two dimensional problem, the governing physical law or principle might be expressed by a partial differential equation (PDE), for example[1]:

$$\frac{\partial^2 T}{\partial x^2} + \frac{\partial^2 T}{\partial y^2} = 0$$

that is valid in the interior of the region R. The solution to the problem must satisfy some boundary conditions or *constraints*, for example $T = T(x,y)$, prescribed on the boundary B. Both interior and exterior boundaries might be present and can be arbitrarily shaped. Note that this governing PDE may be (and usually is!) the result of simplifying assumptions made about the physical system, such as the material being homogeneous and isotropic, with constant linear properties, and so on.

---

[1] The PDE given represents the steady state temperature distribution within a planar body which is governed by the conduction of heat within the body. There are no heat sources, and temperature on the boundary of the body is known.

In order to analyze this problem, the region R is *discretized* (divided) into individual *finite elements* that collectively approximate the shape of the region, as shown in Figure 1(b). This discretization is accomplished by locating *nodes* along the boundary and in the interior of the region. The nodes are then joined by lines to create the finite elements. In 2D problems, these can be triangles or quadrilaterals; in 3D problems, the elements can be tetrahedra or 8-node "bricks". In some FEA software, other higher order types of elements are also possible (e.g. hexagonal prisms). Some higher order elements also have additional nodes along their edges. Collectively, the set of all the elements is called a *finite element mesh*. In the early days of FEA, a great deal of effort was required to set up the mesh. More recently, automatic meshing routines have been developed in order to do most, if not all, of this tedious task.

In the FEA solution, values of the dependent variable (T, in our example) are computed only at the nodes. The variation of the variable within each element is computed from the nodal values so as to approximately satisfy the governing PDE. One way of doing this is by using interpolating polynomials. In order for the PDE to be satisfied, the nodal values of each element must satisfy a set of conditions represented by several linear algebraic equations usually involving other nodal values.

The boundary conditions are implemented by specifying the values of the variables on the boundary nodes. There is no guarantee that the true boundary conditions on the continuous boundary B are satisfied between the nodes on the discretized boundary.

When all the individual elements in the mesh are combined, the discretization and interpolation procedures result in a conversion of the problem from the solution of a continuous differential equation into a very large set of simultaneous linear algebraic equations. This system will typically have many (sometimes hundreds of) thousands of equations in it, requiring special and efficient numerical algorithms. The solution of this algebraic system contains the nodal values that collectively represent an *approximation* to the continuous solution of the initial PDE. An important issue, then, is the accuracy of this approximation. In classical FEA solutions, the approximation becomes more accurate as the mesh is refined with smaller elements. In the limit of zero mesh size, requiring an infinite number of equations, the FEA solution to the PDE would be exact. This is, of course, not achievable. So, a major issue revolves around the question "How fine a mesh is required to produce answers of acceptable accuracy?" and the practical question is "Is it feasible to compute this solution?" We will see a bit later how MECHANICA solves these problems.

IMPORTANT POINT: In FEA stress analysis problems, the dependent variable in the governing PDE's is the displacement from the reference (usually unloaded) position. The material strain (displacement per unit length) is then computed from the displacement by taking the derivative with respect to position. Finally, the stress components at any point in the material are computed from the strain at that point. Thus, if the interpolating polynomial for the spatial variation of the displacement field is linear within an element, then the strain and stress will be constant within that element, since the derivative of a linear function is a constant. The significance of this will be illustrated a bit later in this lesson.

The FEA Model and General Processing Steps

Throughout this manual, we will be using the term "model" extensively. We need to have a clear idea of what we mean by the FEA model.

To get from the "real world" physical problem to the approximate FEA solution, we must go through a number of simplifying steps. At each step, it is necessary to make decisions about what assumptions or simplifications will be required in order to reach a final workable model. By "workable", we mean that the FEA model must allow us to compute the results of interest (for example, the maximum stress in the material) with sufficient accuracy and with available time and resources. It is no good building a model that is over-simplified to the point where it cannot produce the results with sufficient accuracy. It is also no good producing a model that is "perfect" but will not yield useful computational results for several weeks! Quite often, the FEA user must compromise between the two extremes - accepting a slightly less accurate answer in a reasonable solution time. The trick is to know exactly (or even approximately) how much accuracy you are trading off for the reduced processing time. Fortunately, MECHANICA makes this determination a part of the solution itself.

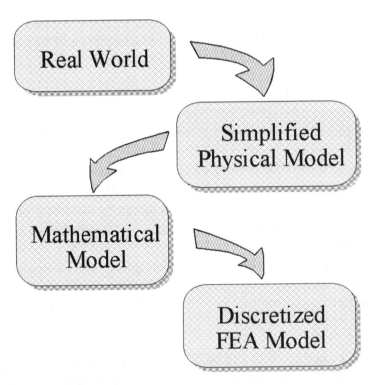

Figure 2 Developing a Model in Finite Element Analysis

To arrive at a model suitable for FEA, we must go through the simplifying steps shown in Figure 2, as follows:

Real World ➜ Simplified Physical Model

This simplification step involves making assumptions about physical properties or the physical layout and geometry of the problem. For example, we usually assume that materials are homogeneous and isotropic and free of internal defects or flaws. It is also common to ignore aspects of the geometry that will have no (anticipated) effect on the results, such as the chamfered and filleted edges on the bracket shown in Figure 3, and perhaps even the mounting holes themselves. Ignoring these "cosmetic" features, as shown in Figure 4, is often necessary in order to reduce the geometric complexity so that the resulting FEA model is practical.

Figure 3 The "Real World" Object

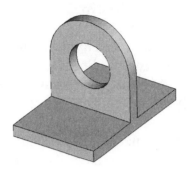

Figure 4 The idealized physical model

Simple Physical Model ➡ Mathematical Model

To arrive at the mathematical model, we make assumptions like linearity of material properties, idealization of loading conditions, and so on, in order to apply our mathematical formulas to complex problems. We often assume that loading is steady, that fixed points are perfectly fixed, beams are long and slender, and so on. As discussed above, the mathematical model usually consists of one or more differential equations that describe the variation of the variable of interest within the boundaries of the model.

Mathematical Model ➡ FEA Model

The simplified geometry of the model is discretized (see Figure 5), so that the governing differential equations obtained in the previous step can be rewritten as a (large) number of simultaneous linear equations representing the assembly of elements in the model.

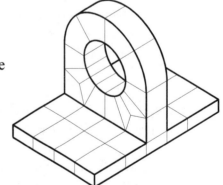

In the operation of FEA software, the three modeling steps described above often appear to be merged. In fact, most of it occurs below the surface (you will never see the governing PDE, for example) or is inherent in the software itself. For example, MECHANICA automatically assumes that materials are homogeneous, isotropic, and linear. However, it is useful to remind yourself about these separate aspects of modeling from time to time, because each is a potential source of error or inaccuracy in the results.

Figure 5 A mesh of solid brick elements

Steps in Preparing an FEA Model for Solution

Starting from the simplified geometric model, there are generally several steps to be followed in the analysis. These are:

1.   identify the model type
2.   specify the material properties, model constraints, and applied loads
3.   discretize the geometry to produce a finite element mesh
4.   solve the system of linear equations
5.   compute items of interest from the solution variables
6.   display and critically review results and, if necessary, repeat the analysis

The overall procedure is illustrated in Figure 6. Some additional detail on each of these steps is given below. The major steps must be executed in order, and each must be done correctly before proceeding to the next step. When a problem is to be re-analyzed (for example, if a stress analysis is to be performed for the same geometry but different loads), it will not usually be necessary to return all the way to the beginning. The available re-entry points will become clear as you move through these tutorials.

The steps shown in the figure are:

1.   The geometric model of the part/system is created using Pro/ENGINEER.

2.   On entry to MECHANICA, the model type must be identified. The default is a solid model.

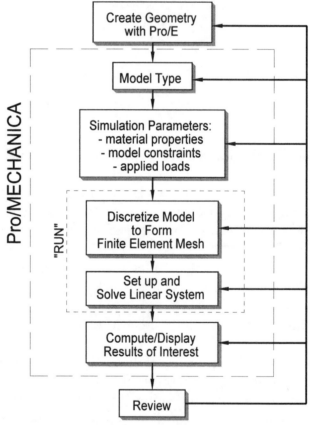

Figure 6 Overall steps in FEA Solution

3.   A)   Specify material properties for the model. It is not necessary that all the elements have the same properties. In an assembly, for example, different parts can be made of different materials. For stress analysis the required properties are Young's modulus and Poisson's ratio. Most FEA packages contain built-in libraries containing properties of common materials (steel, iron, aluminum, etc.).

     B)   Identify the constraints on the solution. In stress analysis, these could be fixed points, points of specified displacement, or points free to move in specified directions only.

     C)   Specify the applied loads on the model (point loads, uniform edge loads, pressure on surfaces, etc.).

4.   Once you are satisfied with your model, you set up and run a processor that actually performs the solution to the posed FEA problem. This starts with the automatic creation of the finite element mesh from the geometric model by a subprogram within MECHANICA called AutoGEM. MECHANICA will trap some modeling errors here, such as elements with no material properties assigned to them. The FEA engine then sets up and solves the system of linear equations. Since this system can be very large (even hundreds of thousands of equations), special numerical techniques are used to exploit certain properties of the system. The processor will produce a summary file of output messages which can be consulted if something goes wrong - for example, a model that is not sufficiently constrained by boundary conditions.

5.   FEA produces immense volumes of output data. The only feasible way of examining this is graphically. MECHANICA has very powerful graphics capabilities to examine the results of the FEA - displaced shape, stress distributions, mode shapes, etc. Hard copy of the results file and screen display is easy to obtain. You can even directly produce web pages with result images and comments.

6.   Finally, the results must be reviewed critically. In the first instance, the results should agree with our modeling intent. For example, if we look at an animated view of the deformation, we can easily see if our boundary constraints have been implemented properly. The results should also satisfy our intuition about the solution (stress concentration around a hole, for example). If there is any cause for concern, it may be advisable to revisit some aspects of the model and perform the analysis again (with slightly different parameters) to explore the model's behavior. This can often identify flaws in how the model was set up[2].

---

[2] You should try to learn something from every run of the model. As stated by Richard Hamming in his book <u>Numerical Methods for Scientists and Engineers</u> (1962): "The purpose of computing is insight, not numbers."

P-Elements versus H-Elements

Not all discretized finite elements are created equal! Here is where a major difference arises between MECHANICA and most other FEA programs.

Convergence of H-elements (the "classic" approach)

Following the classic approach, other programs often use low order interpolating polynomials in each element. This has significant ramifications, especially in stress analysis. As mentioned above, in stress analysis the primary solution variables are the displacements of the nodes. The interpolating functions are typically linear (first order) within each element. Strain is obtained by taking the derivatives of the displacement field and the stress is computed from the material strain. For a first order interpolating polynomial within the element, this means that the strain and therefore the stress components within the element are constant everywhere. The situation is depicted in Figure 7, which shows the computed Von Mises stress in each of the elements surrounding a hole in a thin plate under tension. Such discontinuity in the stress field between elements is, of course, unrealistic and will lead to inaccurate values for the maximum stress. Low order elements lead to the greatest inaccuracy precisely in the regions of greatest interest, typically where there are large gradients within the real object.

An even more disastrous situation is shown in Figure 8. This is a solid cantilever beam modeled using solid brick elements with a uniform transverse load. With only a single first-order element through the thickness, the computed stress will be the same on the top and bottom of the beam. This is clearly wrong, yet the FEA literature and product demonstrations abound with examples similar to this.

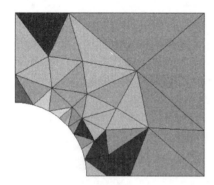

Figure 7 Von Mises stress in 1/4 model of thin plate under tension using first order elements

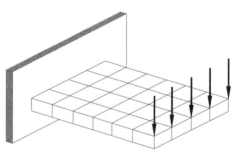

Figure 8 A disaster waiting to happen using first order elements

This situation is often masked by the post-processing capabilities of the software being used, which will sometimes average or interpolate contour values within the mesh or perform other "smoothing" functions strictly for visual appearance. This is strictly a post-processing step, and may bear no resemblance at all to what is actually going on in the mathematical model or (especially) in the real object.

When using first order elements, then, in order to get a more accurate estimate of the stress, it is necessary to use much smaller elements, a process called *mesh refinement*. It may not always be possible to easily identify regions where mesh refinement is required, and quite often the entire mesh is modified. The process of mesh refinement continues until further mesh division and refinement does not lead to significant changes in the obtained solution. The process of continued mesh refinement leading to a "good" solution is called *convergence analysis*. Of course, in the process of mesh refinement, the size of the computational problem becomes larger and larger and we may reach a limit for practical problems (due to time and/or memory limits) before we have successfully converged to an acceptable solution.

The use of mesh refinement for convergence analysis leads to the *h-element* class of FEA methods. This "h" is borrowed from the field of numerical analysis, where it denotes the fact that convergence and accuracy are related (sometimes proportional to) the step size used in the solution, usually denoted by *h*. In FEA, the *h* refers to the size of the elements. The elements, always of low order, are referred to as h-elements, and the mesh refinement procedure is called *h-convergence*. This situation is depicted in parts (a) and (b) of Figure 9, where a series of constant-height steps is used to approximate a smooth continuous function. The narrower the steps, the more closely we can approximate the smooth function. Note also that where the gradient of the function is large (such as near the left edge of the figure), then mesh refinement will always produce increasingly higher maximum values.

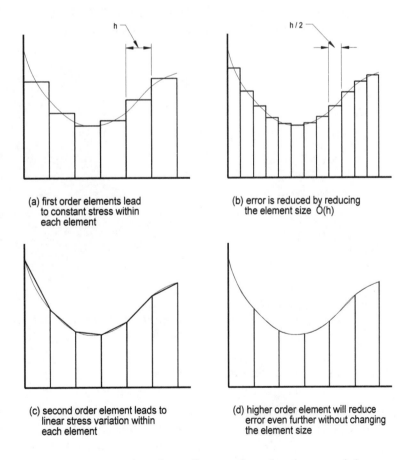

Figure 9 Approximation of stress function in a model

The major outcome of using h-elements is the need for meshes of relatively small elements. Furthermore, h-elements are not very tolerant of shape extremes in terms of skewness, rapid size variation through the mesh, large aspect ratio, and so on. This further increases the number of elements required for an acceptable mesh, and this, of course, greatly increases the computational cost of the solution.

Convergence of P-elements (the MECHANICA approach)

Now, the major difference incorporated in MECHANICA is the following: instead of constantly refining and recreating finer and finer meshes, convergence is obtained by *increasing the order of the interpolating polynomials on each element*. The mesh stays the same for every iteration, called a *p-loop pass*. The use of higher order interpolating polynomials for convergence analysis leads to the *p-element* class of FEA methods, where the "p" denotes polynomial. This method is depicted in parts (c) and (d) of the Figure 9. Only elements in regions of high gradients are bumped up to higher order polynomials. Furthermore, by examining the effects of going to higher order polynomials, MECHANICA can monitor the expected error in the solution, and automatically increase the polynomial order only on those elements were it is required. Thus, the convergence analysis is performed quite automatically, with the solution proceeding until an accuracy limit (set by the user) has been satisfied. With MECHANICA, the limit for the polynomial order is 9. In theory, it would be possible to go to higher orders than this, but the computational cost starts to rise too quickly. If the solution cannot converge even with these 9th order polynomials, it may be necessary to recreate the mesh at a slightly higher density so that lower order polynomials will be sufficient. This is a very rare occurrence.

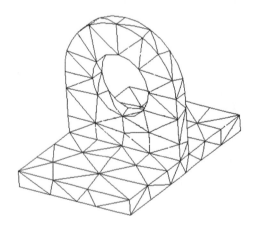

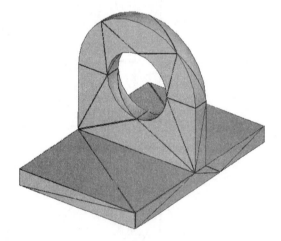

Figure 10 A mesh of solid tetrahedral (4 node) h-elements

Figure 11 A mesh of tetrahedral p-elements produced by MECHANICA.

The use of p-elements has a number of features/advantages:

▸   The same mesh can be used throughout the convergence analysis, rather than recreating meshes or local mesh refinement required by h-codes. For complicated geometries, eliminating the need for re-meshing results in significantly less CPU time.

▸   The mesh is virtually always more coarse and contains fewer elements than h-codes. Compare the meshes in Figures 10 and 11, and note that the mesh of h-elements in Figure

10 would probably not produce very good results, depending on the loads and constraints applied. The reduced number of elements in MECHANICA (which can be a couple of orders of magnitude smaller) initially reduces the computational load, but as the order of the polynomials gets higher, this advantage is somewhat diminished.

▸ The restrictions on element size and shape are not nearly as stringent for p-elements as they are for h-elements (where concerns of aspect ratio, skewness, and so on often arise).

▸ Automatic mesh generators, which sometimes produce very poor meshes for h-elements, are much more effective with p-elements, due to the reduced requirements and limitations on mesh geometry.

▸ Since the same mesh is used throughout the analysis, this mesh can be tied directly to the geometry. This is the key reason why MECHANICA is able to perform sensitivity and optimization studies during which the geometric parameters of a body can change, but the program does not need to be constantly re-meshing the part.

Convergence and Accuracy in the Solution

It should be apparent that, due to the number of simplifying assumptions necessary to obtain results with FEA, we should be quite cautious about the results obtained. No FEA solution should be accepted unless the convergence properties have been examined.

For h-elements, this generally means doing the problem several times with successively smaller elements and monitoring the change in the solutions. When decreasing the element size results in a negligible (or acceptably small) change in the solution, then we are generally satisfied that the FEA has wrung all the information out of the model that it can.

As mentioned above, with p-elements, the convergence analysis is built in to the program. Since the geometry of the mesh does not change, no remeshing is required. Rather, each successive solution (called a *p-loop pass*) is performed with increasing orders of polynomials (only on elements where this is required) until the change between iterations is "small enough". Figure 12 shows the convergence behavior of two common measures used to monitor convergence in MECHANICA. These are the maximum Von Mises stress and the total strain energy. Note that the Von Mises stress will generally always increase during the convergence test, but can behave quite erratically as we will see later. Because Von Mises stress is a local measure, the strain energy is probably a better measure to use to monitor convergence.

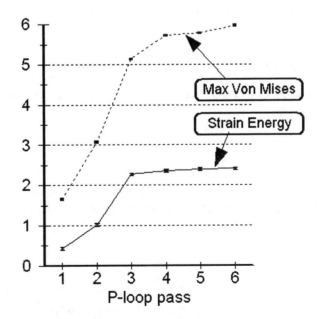

Figure 12 Two common convergence measures using p-elements.

Sources of Error

Error enters into the FEA process in a number of ways:

♦ **errors in problem definition** - are the geometry, loads, and constraints known and implemented accurately? Is the correct analysis being performed? Are the material properties correct and/or appropriate?

♦ **errors in creating the physical model** - can we really use symmetry? Is the material isotropic and homogeneous, as assumed? Are the physical constants known? Does the material behave linearly?

♦ **errors in creating the mathematical model** - is the model complete enough to capture the effects we wish to observe? Is the model overly complex? Does the mathematical model correctly express the physics of the problem?

♦ **errors in discretization** - is the mesh too coarse or too fine? Have we left accidental "holes" in the model? If using shell elements, are there tears or rips (free edges) between elements where there shouldn't be?

♦ **errors in the numerical solution** - when dealing with very large computational problems, we must always be concerned about the effects of accumulated round-off error. Can this error be estimated? How trustworthy is the answer going to be?

♦ **errors in interpretation of the results** - are we looking at the results in the right way to see what we want and need to see? Are the limitations of the program understood[3]? Has the possible misuse of a purely graphical or display tool obscured, hidden, or misrepresented a critical result?

You will be able to answer most of these questions by the time you complete this tutorial. The answers to others will be problem dependent and will require some experience and further exposure before you are a confident and competent FEA user.

A CAD Model is *NOT* an FEA Model!

One of the common misconceptions within the engineering community is the equivalence of a CAD solid model with a model used for FEA. These are *not* the same despite proclamations of the CAD vendors that their solid models can be "seamlessly" ported to one or another FEA program. In fact, this is probably quite undesirable! It should not be surprising that CAD and FEA models are different, since the two models are developed for different purposes.

---

[3] The author once had a student who was rightly concerned about the very large deflections in a truss computed using a simple FEA program. The program was performing a linear analysis, and was computing stresses in some members two orders of magnitude higher than the yield strength of the material. The student did not realize that the software knew nothing about failure of the material. It turned out that a simple data entry error had reduced the cross sectional area of the members in the truss.

The CAD model is usually developed to provide a data base for manufacturing. Thus, dimensions must be fully specified (including tolerances), all minor features (such as fillets, rounds, holes) must be included, processing steps and surface finishes are indicated, threads are specified, and so on. Figure 13 shows a CAD solid model of a hypothetical piping component, complete with bolt holes, flanges, o-ring grooves, chamfered edges, and carrying lugs. Not visible in the figure are the dimensions, tolerances, and welding instructions for fabrication which are all part of the CAD model.

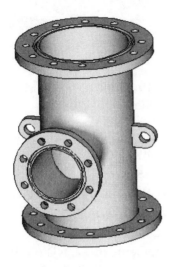

Figure 13 A hypothetical 3D solid model of a piping junction

Figure 14 The 3D solid model of pipe junction

FEA is usually directed at finding out other information about a proposed design. To do this *efficiently*, the FEA model can (and often needs to) be quite different from the CAD model. A simple example of this is that the symmetry of an object is often exploited in the preparation of the FEA model. In one of the exercises we will do later, we will model a thin tapered plate with a couple of large holes. The plate has a plane of symmetry so that we only need to do FEA of one-half of the plate. It is also quite common in FEA to ignore minor features like rounds, fillets, chamfers, holes, minor changes in surface profile, and other cosmetic features unless these features will have a large effect on the measures of interest in the model. Most frequently, they do not, and can be ignored.

Figure 14 shows an FEA model of the piping component created to determine the maximum Von Mises stress in the vicinity of the filleted connection between the two pipes. The differences between the two models shown in Figures 13 and 14 are immediately obvious. Figures 15 and 16 show the mesh of shell elements created from the surface, and the computed Von Mises stress.

Figure 15 Shell elements of specified thickness created from 3D model

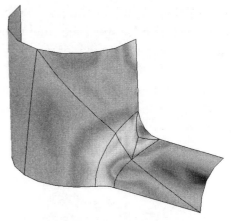

Figure 16 Von Mises stress in the FEA model

In summary, the stated goal of FEA (the "Golden Rule", if you like) might be expressed as:

> *Use the simplest model possible that will yield sufficiently reliable results of interest at the lowest computational cost.*

You can easily see how this might be at odds with the requirements of a CAD model. For further discussion of this, see the excellent book **Building Better Products with Finite Element Analysis** by Vince Adams and Abraham Askenazi, Onword Press, 1998.

Overview of MECHANICA Structure

Basic Operation

We are going to start using MECHANICA in the next chapter. Before we dive in, it will be useful to have an overall look at the function and organization of the software. This will help to explain some of the MECHANICA terminology and see how the program relates to the ideas presented in this chapter's overview of FEA.

We can divide the operation and functionality of MECHANICA Structure according to the rows in Table I below. These entries are further elaborated in the next few pages. In the process of setting up and running a solution, you will basically need to pick one option from each row in the table. The top-down organization of the table is roughly in the order that these decisions must be made. Other issues such as creation of the model geometry and post-processing and display of final results will be left to subsequent chapters.

TABLE I - An Overall View of MECHANICA Capability and Function

	MECHANICA Options	Description
Mode of Operation	Independent Integrated	how MECHANICA is operated with respect to Pro/ENGINEER
Type of Model	3D Plane Stress Plane Strain Axisymmetric	basic structure of the model
Type of Elements	Shell Beam Solid Spring Mass	element types that can be used in a model

Analysis Methods	**Structure**:
	  Static
	  Modal
	  Buckling
	  Pre-stress modal
	  Pre-stress buckling
	Thermal:
	  Steady state
	  Transient
		the fundamental solution being sought for the model

Convergence Methods	Quick Check Single Pass Adaptive Multi-Pass Adaptive	method of monitoring convergence in the solution
Design Studies	Standard Sensitivity Optimization	high level methods to organize essentially repetitive computations

Modes of Operation

A discussion of the full details of operating modes gets pretty confusing, so only the main points are presented here. These are:

1.   MECHANICA can operate in two modes[4], in relation to its cousin application Pro/ENGINEER. These are: **independent** and **integrated**. A special license is required to run the independent version. In the student edition, only integrated mode is possible.

---

[4] A third mode, called *linked*, was available up until Release 2000*i*, but has been removed.

2. The user interface is determined by the mode:
   - ♦ integrated mode - Pro/ENGINEER interface
   - ♦ independent mode - MECHANICA interface
3. If you start out in Pro/ENGINEER to create the part (or assembly) geometry and call up MECHANICA, you will initially be running in integrated mode. You can then switch to independent mode if desired (and if your license allows it), as illustrated here (note that the arrow is a one-way transfer - you can't get back again!):

   Integrated ➜ Independent
   Mode Mode

4. If you switch to independent mode, the connection with Pro/ENGINEER will be severed. Any changes in design parameters (for example following an optimization) must be manually transferred back into the Pro/E model.
5. In integrated mode, a few MECHANICA commands and result displays may not be available. However the tight integration with Pro/E makes it very easy to perform design modification and quick FEA.
6. In integrated mode, the user interface is the same as Pro/E. Only one set of controls to learn! The independent mode user interface is quite different.
7. The full set of MECHANICA commands and functions are available in independent mode (for example: manual and semi-automatic mesh generation for difficult models, excluding elements from convergence monitoring).
8. Although independent mode gives access to the complete range of MECHANICA functionality, the benefits of feature-based geometry creation/modification are lost.

A condensed comparison of these operating modes is shown in Table II. As mentioned above, all the tutorials in this manual are meant to be run in integrated mode.

TABLE II - MECHANICA Modes of Operation

Integrated Mode	Independent Mode
Pro/E interface	MECHANICA interface
all analyses available	all analyses available
2D and 3D models	2D and 3D models
some measures of results not available	all measures available
some analysis options not available (eg excluding elements)	all options available
all elements generated automatically (but with some mesh controls)	element creation manual, semi-, or fully automatic
sensitivity and optimization using Pro/E parameters only	sensitivity and optimization uses MECHANICA variables

Types of Models

This is fairly self-explanatory. In addition to 3D solid, shell, and beam models, MECHANICA in both modes can treat 2D models (plane stress, plane strain, or axisymmetric). Note that all geometry and model entities (loads and constraints) for all 2D model types must be defined in the XY plane of a selected coordinate system. Also, a very thin plate might be modeled as a 2D shell, but if it is loaded with any force components normal to the plate, then it becomes a 3D problem.

Independent MECHANICA contains a good set of tools to create both 2D and 3D geometry. However, these are neither feature-based nor parametric. Complicated 3D geometry of parts would be easier to make in Pro/E or some other CAD package, and brought into MECHANICA in integrated mode. The model geometry is generally created entirely in Pro/E. It is possible to create some (non-solid) simulation features while in MECHANICA, such as datum points and curves.

Types of Elements

The various types of elements that can be used in MECHANICA are listed in Table I. It is possible to use different types of elements in the same model (e.g. combining solid + beam + spring elements), but we will discuss only a couple of models of this degree of complexity in these tutorials. At first glance, this seems like a limited list of element types. H-element programs typically have large libraries of different element types (sometimes several dozen), but these are often necessary to overcome the limitations of low order simple h-elements. In MECHANICA, we do not have this problem and you can do practically anything with the elements available.

Analysis Methods

For a given model, several different analysis types are possible. For example, in *Structure* the *static* analysis will compute the stresses and deformations within the model, while the *modal* analysis will compute the mode shapes and natural frequencies. *Buckling* analysis will compute the buckling loads on the body, and so. Other analysis methods are available in *Structure* but in this manual, we will only look at static stress and modal analysis. In *Thermal*, the two analysis types are to determine either the steady state temperature distribution, or the transient response of a model given temperature or heat flux boundary conditions.

Convergence Methods

As discussed above, using the p-code method allows MECHANICA to monitor the solution and modify the polynomial edge order until a solution has been achieved to a specified accuracy. This is implemented with three options:

- **Quick Check** - This actually isn't a convergence method since the model is run only for a single fixed (low, usually 3) polynomial order. **The results of a Quick Check should never be trusted.** What a Quick Check is for is to quickly run the model through the solver in order to pick up any errors that may have been made, for example in the constraints. A quick review of the results will also indicate whether any gross modeling errors have been made and possibly point out potential problem areas in the model.

- **Single Pass Adaptive (SPA)** - More than a Quick Check, but less than a complete convergence run, the single pass adaptive method performs one pass at a low polynomial order, assesses the accuracy of the solution, modifies the p-level of "problem elements", and does a final pass with some elements raised to an order that should provide reasonable results. If the model is very computationally intensive and/or is very well behaved and understood, this can save you a lot of time. The Single Pass Adaptive analysis is the recommended option for most model types. If you suspect problems in the model, or want further verification of convergence, you might want to pick the next option, MPA.

- **Multi-Pass Adaptive (MPA)** - The ultimate in convergence analysis. Multiple "p-loop" passes are made through the solver, with edge orders of "problem elements" being increased by one or two with each pass. This iterative approach continues until either the solution converges to a specified accuracy or the maximum specified edge order (default 6, maximum 9) is reached. Convergence is monitored by watching one or more "measures", or numerical aspects of the solution. Typical measures are the maximum Von Mises stress, or strain energy (as seen in Figure 12). The default measures for a static stress analysis are nodal displacement, local strain energy, and global RMS (root-mean-square) stress. At the conclusion of the run, the convergence measures may be examined. Although it is not strictly necessary, in this manual we will use mostly the MPA option to get a feel for how the analysis is performing.

Design Studies

A Design Study is a problem or set of problems that you define for a particular model. When you ultimately press the **Run** button on MECHANICA, what will execute is a design study - it is the top-most level of organization in MECHANICA. There are three types of design studies:

- A **Standard** design study is the most basic and simple. It will include at least one but possibly several analyses (for example a static analysis plus a modal analysis). For this study, you need to specify the geometry, create the elements, assign material properties, set up loads and constraints, determine the analysis and convergence types, and then display and review the final results. The Standard design study is what most people would consider "Finite Element Analysis."
- A **sensitivity** design study can be set up so that results are computed for several different values of designated design variables or material properties. In addition to the standard model, you need to designate the design variables and the range over which you want them to vary. You can use a sensitivity study to determine, for example, which design variables will have the most effect on a particular measure of performance of the design like the maximum stress or total mass.
- Finally, the most powerful design study is an **optimization**. For this, you start with a basic FEA model. You then specify a desired goal (such as minimum mass of the body), geometric constraints (such as dimensions or locations of geometric entities), material constraints (such as maximum allowed stress) and one or more design variables which can vary over specified ranges. MECHANICA will then search through the space of the design variables and determine the best design that satisfies your constraints. Amazing!

A Brief Note about Units

It is crucial to use a consistent set of units throughout your MECHANICA activities. The program itself has no default set of units (other than those brought in with the model from Pro/E), and only uses the numerical values provided by you. Thus, if your geometry is created with a particular linear unit like mm or inches in mind, you must make sure that any other data supplied, such as loads (force, pressure) and material properties (density, Young's modulus, and so on) are defined consistently. The built-in material libraries offer properties for common materials in four sets of units (all at room temperature):

> inch - pound - second
> foot - pound - second
> meter - Newton - second
> millimeter - Newton - second

Note that the weight of the material is obtained by multiplying the mass density property by the acceleration of gravity expressed in the appropriate unit system.

If you require or wish to use a different system of units, you can enter your own material properties, but must look after consistency yourself. Table III outlines the common units in the various systems including how some common results will be reported by MECHANICA. For further information on units, consult the on-line help pages "About Units" and "Unit Conversion Tables."

TABLE III - Common unit systems in MECHANICA

Quantity	System and Units			
	SI MNS	Metric mm-N-s	English FPS ft-lb-sec	English IPS in-lb-sec
length	m	mm	ft	in
time	s	s	sec	sec
mass	kg	tonne (1000 kg)	slug	$lbf\text{-}sec^2 / in$
density	kg/m^3	$tonne/mm^3$	$slug/ft^3$	$lbf\text{-}sec^2 / in^4$
gravity, g	$9.81 \ m/s^2$	$9810 \ mm/s^2$	$32.2 \ ft/sec^2$	$386.4 \ in/sec^2$
force	N	N	lbf	lbf
stress, pressure, Young's modulus	$N/m^2 = Pa$	$N/mm^2 = MPa$	lbf/ft^2	$lbf/in^2 = psi$

The units systems for *Thermal* are particularly important (and confusing!). We will leave those until we need them in Chapter 10.

Files and Directories Produced by MECHANICA

Since you will be working in integrated mode in this book, note that your entire simulation model is stored in the Pro/E part file. You do not need to store a special copy of this. Simulation entities like loads and constraints (plus all other data and settings you define for the model) will appear when you transfer into MECHANICA from Pro/E.

MECHANICA produces a bewildering array of files and directories. Unless you specify otherwise (or have specified in your default system configuration), all of these will be created in the Pro/E working directory. It is therefore wise to create a new subdirectory for each model, make it your working directory, and store the part file there. Locations for temporary and output files can be changed at appropriate points in the program. For example, when you set up to run a design study, you can designate the location for the subdirectory which MECHANICA will create for writing the output files.

The important files and directories are indicated in the Table IV. In the table, the symbol ① represents the directory specified in the *Run Settings* dialog box for output files, and ② represents the directory specified in the same dialog box for temporary files. Unless the run terminates abnormally, all temporary files are deleted on completion of a run. The names *model*, *study*, and *filename* are supplied by you during execution of the program. Note that many of these files are stored in a binary format and are not readable by normal file editors.

On-line Documentation

For further details on any of these functions or operating commands, consult the on-line documentation available with MECHANICA. See your local system administrator for information on how to access these files. Basically, you select

Help > Help Center

in the pull-down menu. Wildfire's Help pages are viewed using a built-in browser. Select the **Contents** tab and expand the list to see an overview of help topics (Figure 17). Also, check out the Index and Search functions.

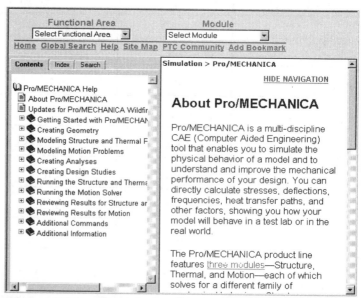

Figure 17 The *Help Center* main page

Table IV - Some Files Produced by MECHANICA

File Type	File/Directory Name	Comments
Model Files	*model.***mdb** *model.***mbk**	the **mdb** file contains the last-saved model database. **mbk** is a backup that can be used if the **mdb** file is lost or corrupted
Engine Files	①*/study/study.***mdb**	contains the entire model database at the time a design study is started
	①*/study/study.***cnv** ①*/study/study.***hst** ①*/study/study.***opt** ①*/study/study.***res** ①*/study/study.***rpt**	Engine output files: - convergence information - model updates during optimization - optimization data - measures at each pass - output report for a design study (also accessible with the ***Study Status*** command)
Exchange Files	*filename.***dxf** *filename.***igs**	file formats used for import/export of geometry information
Temporary Files	②*/study.***tmp/*.tmp** ②*/study.***tmp/*.bas**	should delete automatically on completion of design study
Results Files	*filename.***rwd** *filename.***grt**	- result window definitions stored with Save in the Result Windows dialog box - graph data report
AutoGEM Files	*model.***agm**	information about the most recent AutoGEM operation. If the model has not yet been named, this file is **untitled.agm**
Miscellaneous Files	**mech_trl.txt.?**	a complete history of the MECHANICA session (every command, mouse click, and data entry). Automatically incremented with each new session (same as a Pro/E trail file).

Summary

This chapter has introduced the background to FEA. In particular, the differences between h-code and p-code methods have been discussed. The general procedure involved in performing an analysis was described. Finally, an overview of MECHANICA has been presented to give you a view of the forest before we start looking at the individual trees!

You are strongly urged to have a look at the articles written by Dr. Paul Kurowski that are listed in the References at the end of this chapter. These offer an in-depth look at common errors made in FEA, the concept of convergence, a comparison of h- and p-elements, and more comments on the difference between CAD and FEA.

In the next Chapter, we will start to look at the basic tools within MECHANICA. We will produce a simple model and go through the process of setting up a standard design study for static analysis of a simple 3D solid model. We will also take a first look at the methods for viewing the results of the analysis.

References

The following articles by Dr. Kurowski are available on-line at **www.machinedesign.com**:

> "*Avoiding Pitfalls in FEA*," Paul Kurowski, *Machine Design*, November 1994.
> "*When good engineers deliver bad FEA*," Paul Kurowski, *Machine Design*, November, 1995.
> "*Good Solid Modeling, Bad FEA*," Paul Kurowski, *Machine Design*, November, 1996.
> "*How to find errors in finite-element models*," Paul Kurowski & Barna Szabo, *Machine Design*, September, 1997.
> "*Easily made errors that mar FEA results*," Paul Kurowski, *Machine Design*, September, 2001.
> "*More errors that mar FEA results*," Paul Kurowski, *Machine Design*, March, 2002.

Finite Element Methods for Engineers, Roger T. Fenner, Macmillan, 1975.

Building Better Products with Finite Element Analysis, Vince Adams and Abraham Askenazi, Onword Press, 1998.

The Finite Element Method in Mechanical Design, Charles E. Knight, Jr., PWS-Kent, 1993.

CAD/CAM Theory and Practice, Ibrahim Zeid, McGraw-Hill, 1991.

The Finite Element Method, T.J.R. Hughes, Prentice Hall, 1987.

Computer-Assisted Mechanical Design, J.Ed Akin, Prentice Hall, 1990.

Questions for Review

1.   What is the purpose of interpolating polynomials in FEA?
2.   What is a "model?" What are some different types of models and how do these relate to the real world?
3.   Is it ever possible for a FEA solution to be "exact?" Why or why not?
4.   What is the primary source of error when using first-order h-elements for stress analysis?
5.   Give an outline of the necessary steps in performing FEA.
6.   Why is it probably not a good idea to use a CAD model directly in an FEA solution?
7.   What is the "Golden Rule" of FEA?
8.   How is convergence of the solution obtained using h-code and p-code methods?
9.   Does mesh refinement always yield higher maximum stresses?
10.   What is the maximum edge order available in MECHANICA? In the (unlikely) event that the solution will not converge, what needs to be done?
11.   What measures are typically used in MECHANICA to monitor convergence?
12.   How will error enter into an FEA?
13.   What is a *design study*? What types are available in MECHANICA? How is it different from an *analysis*?
14.   What are the three methods of convergence analysis? When would each be appropriate?
15.   What types of 2D models can be created? In what operating modes? What restrictions are there on 2D models?
16.   What types of analyses can be performed on a model?
17.   How can you gain access to the on-line help on your system?
18.   Compare the advantages and disadvantages of *integrated* and *independent* modes of operation.
19.   Is specifying the convergence in MECHANICA considered to be part of the model, analysis, or design study?
20.   Where and how do you set up the units for the MECHANICA model?

Exercises

1.   Consult a numerical methods textbook and find out what algorithms are used to solve very large linear systems. What effect does round-off error have, and can this be quantified? Are some methods more susceptible to round-off than others?

2.   Locate some product brochures for FEM software, and look for the kind of modeling errors discussed in this chapter. Compare the models to the "real thing" and comment on any differences you notice.

This page left blank.

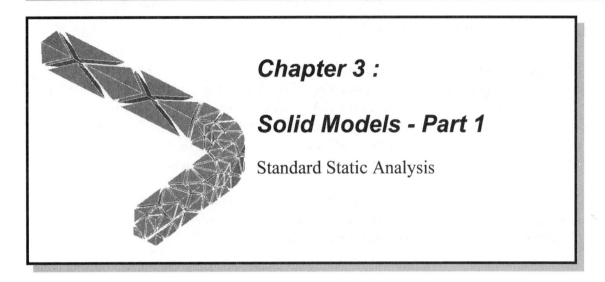

Chapter 3 :

Solid Models - Part 1

Standard Static Analysis

Synopsis

Static analysis of solid models; setting up the model (constraints, loads, material); examining convergence; displaying results; display options and controls; examining the FEA mesh

Overview of this Lesson

In this lesson, we will create a very simple solid model and perform a static stress analysis. This will illustrate the common steps involved in all FEA procedures, as outlined in the previous chapter. Along the way, we will encounter the main dialog windows used throughout MECHANICA for setting options for program operation. In this lesson, we will explore some (not all!) of these options. The main steps involved are:

- creation of the model (in Pro/E)
- specifying model type
- setting up constraints, loads, and material definitions
- defining the analysis and design study
- running the solution
- setting up and showing result displays

After setting up and running this first simple FEA model, we will look at some variations of the finite element mesh and the effect on the results.

There are some Questions for Review and Exercises at the end of the chapter, which will give you some practice with the concepts covered and introduce you to some additional options.

Simple Static Analysis of a Solid Part

Solid models are the default type when working in integrated mode with Pro/ENGINEER. When (most) people think of FEA, they are thinking of or visualizing solid models. These can be treated with a minimum of work within MECHANICA and so are a good place to start. You should note, however, that the majority of this tutorial deals with other types of models. Solids are useful for some types of models, but we will learn in later chapters other techniques that are considerably more efficient for some geometries (beams, shells, plates, and so on).

The FEA mesh for a solid model is composed of tetrahedral elements. These are like a three sided pyramid; each element has 4 faces (including the bottom), hence forms a tetrahedron. The edges and surfaces of each "tet" element can be straight or curved[1].

Creating the Geometry of the Model

The model we are going to study is a simple L-shaped bar with a square cross section, shown in Figure 1. The bar will be fixed (cantilevered) at the left end, and a uniformly distributed force will be applied to the square face at the other end. We are going to be using this model for the next couple of chapters, so set it up exactly as described below. Also, you may want to think a bit about file management here. For example, because of the number of files created during a run, it is common practice to keep each model in its own directory to reduce confusion.

Assuming you have Wildfire running, open the **Folder Navigator** and select the location where you want to put your model directory (maybe in a directory under the current working directory). Click on the current directory and using the RMB (right mouse button) pop-up menu select

> ### *New Folder*

Call the folder something like **CHAP3**. Then, again using the RMB pop-up, select

> ### *Make Working Directory*

Now create a new solid part (or download this one from <**www.schroff1.com**>). We will use units of mm-N-s (millimeter-Newton-second), which are available with one of the part templates (select the template **mmNs_part_solid**). Call the part *[bar01]*.

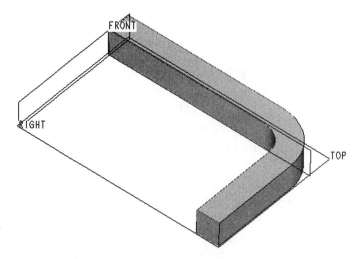

Figure 1 Solid model of L-shaped bar

---

[1] In independent mode, brick and wedge elements are also available. We will see an example of brick elements towards the end of this lesson.

The part template has default datum planes and a coordinate system in place. Our part consists of a single feature - a sweep. Although there are alternative ways of making this model, this is probably the simplest. There are a couple of ways to make this sweep[2]. To create it using a pre-Wildfire command sequence, you can use *Insert > Sweep > Protrusion* and follow the prompts. To create it in a more Wildfire-oriented sequence, first create a sketched datum curve on the **TOP** datum plane - this will become the trajectory of the sweep. The sketching references are **FRONT** and **RIGHT**. The curve is shown in Figure 2. Notice the dimensioning scheme, which will be important to us in the next chapter. When the curve is complete, make sure it is selected (highlighted in red), and select the *Variable Section Sweep* tool in the right toolbar. Observe (and set if necessary) the location of the start point (at the left end in Figure 2). Use Sketcher to create the square cross section shown in Figure 3 (isometric view), noting that the section will be on the outside of the curved corner - this will also be important later. Make sure you create a solid instead of a surface (the default for variable section sweeps). You can also set the sweep option to **Constant Section**.

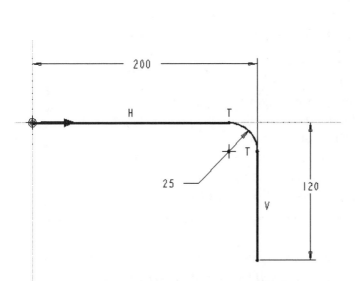

Figure 2 Sketched datum curve that will form the trajectory for the sweep

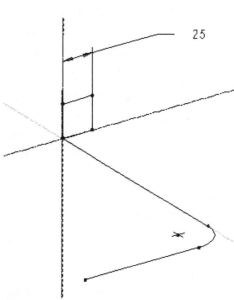

Figure 3 Sweep section (25mm square)

---

[2] If you are not familiar with sweeps, see Lesson #11 in the Pro/ENGINEER Tutorial, or browse the on-line help.

Setting up the FEA Model

Launching MECHANICA

With the geometry specified, we are ready to enter MECHANICA . In the pull-down menus at the top, select:

> *Applications > Mechanica*

An information window appears reminding you what units are being used for the part. These should be millimeters, Newtons, and so on (mass in tonnes!). It is a common "oops" to not have the desired units, so this is a useful reminder. There is a check box at the bottom to turn off the info window if you don't want it next time in the current session. This is not advised so leave this unchecked and select *Continue*.

It takes a few seconds for the program to be launched. The **Model Type** window will open at the right. The two modes are **Structure** (default) and **Thermal**.

> The check box *FEM Mode* allows you to do model pre-processing *only*. With this checked you can apply loads, constraints, define materials, and create the mesh, BUT you cannot run the solver. This mode of operation allows you to create models and produce correctly formatted input files for other commercial FEA solver programs. This is handy because some solvers do not have very easy-to-use preprocessor interfaces, so you can use Wildfire to set up the model much easier. **If you are going to use the MECHANICA solver, *FEM Mode* must be left**

Click the *Advanced* button. This shows the other model types we will be investigating through these lessons. The default type, as mentioned above, is a 3D solid. Select *OK*.

The **Model Type** window will not appear the next time you enter this model from Pro/E. If you want to change the model type in the future, you must select

> *Edit > Mechanica Model Type*

The part display is essentially the same as the Pro/E display. You can turn off the datum plane display, as these will not be needed any further. You can also close the Navigator window. Notice the addition of a green coordinate system labeled **WCS**. This is the World Coordinate System, created by default by MECHANICA . An FEA model can have several coordinate systems (cartesian, cylindrical, or spherical). The currently active system is highlighted in green. Note the directions of the X, Y, and Z axes - these are important to specify directions of loads and constraints.

New toolbars also appear at the top and right. The right toolbar, Figure 4, contains shortcuts to create simulation entities like constraints, loads, and idealized elements. These commands are also available in the top pull-down menus (check out *Insert*, *Properties*, *AutoGEM*, *Analysis*).

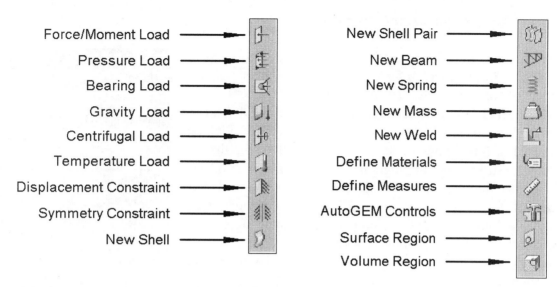

Figure 4 Right toolbar buttons in MECHANICA

You will note several new buttons in the top toolbar. From left to right, these are the following:

♦ The ***Create Mesh*** button is used to create a mesh without actually launching the solver. This would allow you to experiment with mesh control settings.

♦ The ***Design Study*** button which is used to create analysis and design study definitions and then run them.

♦ The ***Review Results*** button gives you access to commands for post-processing (animations, fringe plots, graphs, etc.).

♦ Finally, the last new button is the ***Setup Simulation Display*** tool. This is a shortcut to tools for configuring the appearance of the model on the screen. We'll also explore this a bit later.

Check the preselection filter options in the pull-up list at the bottom right of the screen. This filter is to help you when selecting simulation entities on the screen, and works in the same way as the preselection filter in Pro/E. This function, introduced in Release 2001, is part of the new object-action command structure. We'll explore this a bit more later on.

Note that the datum creation toolbar is still with us. This indicates that it is possible to create these features while we are in MECHANICA if the need arises. If we do, they will be saved with the model but only active (and visible in the model tree) when we are doing the FEA.

Applying the Constraints

We are going to constrain the left end of the bar against all motion, since the bar is cantilevered out from a support. This constraint applies to the entire square surface at the end. In the right toolbar, select

New Displacement Constraint

A new dialog window opens (Figure 5). In the top data field, enter a name *[fixed_face]*. Individual constraints are stored as members of constraint sets. The default name of the set is **ConstraintSet1**. In the **References** list, note that the default is *Surfaces.* We will see later why this is a useful default! Pick on the desired surface on the model. You may want to spin the model. The surface will highlight in red. Then *OK* or middle click. Notice that the default coordinate system is WCS. If you wanted a different system (if it existed!), you could select that here. At the bottom of the Constraint window, there are six possible constraints - three translation and three rotation constraints. Each constraint has four buttons. The buttons (from left to right) are for

- Free
- Fixed
- Prescribed (specified displacement)
- Function of coordinates

The default setting is **Fixed** for all constraints. This is exactly what we want here for the translation of this surface. Any element nodes that lie on this surface will be prevented from moving in any direction.

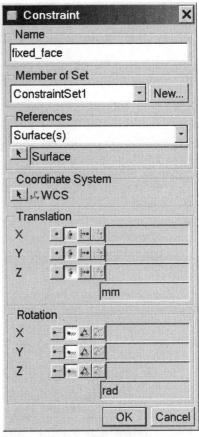

Figure 5 Displacement constraint dialog window

IMPORTANT POINT

Rotational constraints are irrelevant for solid models. The FEA solution of a model using solid elements involves only the translational degrees of freedom of the nodes. Thus, rotation of a node can be neither specified nor computed! These constraints are not inactive (grayed out) at this time because MECHANICA does not know if we are going to be attaching other model entities to this surface (like beams or shells) which can involve rotational constraints.

The completed constraint dialog window is shown in Figure 5. Select *OK* or middle click. A pattern of small red triangles appears on the surface. A large triangular icon appears (Figure 6). This is the constraint icon. The constraints are indicated by the fill pattern of the six boxes across the bottom of this icon. These constraints are currently all fixed, so all boxes are filled in.

Now on to applying the load on the end face.

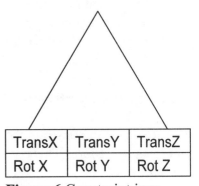

| TransX | TransY | TransZ |
| Rot X | Rot Y | Rot Z |

Figure 6 Constraint icon

Applying the Loads

In the right toolbar, select

New Force/Moment Load

This brings up the dialog window shown in Figure 7. Enter a name *[endload]*. The load is a member of **LoadSet1**. Once again the default reference is **Surfaces**. Click on the square surface at the free end of the bar and middle click. The load will be defined relative to WCS. Select the *Advanced* button, and check out the options available for the load distribution. Select *Total Load* and *Uniform*. The areas at the bottom of the dialog window are for entering the applied forces and moments on the designated surface. These can either use components or (see the pull-down list) a magnitude and direction. Select *Components* and enter the values shown in Figure 7. Note that the units are displayed. Recall the directions of the WCS. Click the *Preview* button to see the applied load on the model. Finally, select *OK*.

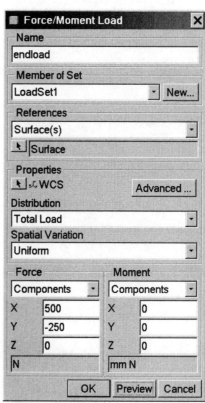

Figure 7 Force/Moment dialog

The model should now appear similar to Figure 8. Note the single bold up-arrow icon that indicates the load is not in the direction of the applied load.

Let's explore some view options for the model. In the pull-down menus, select

View > Simulation Display > Settings

(or use the new button in the top toolbar). This brings up the dialog window shown in Figure 9 which allows us to control the appearance of the display. Experiment with the various options by checking them and using the *Preview* button. For example, try out the *Display Names* and *Load Value*

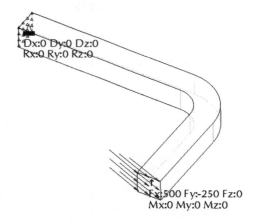

Figure 8 Model with constraint and load

options for the icons. These displays are handy for documenting models - a hard copy of this display is a useful image to archive. Change the Load Arrows display to *Arrow Tails Touching*. When your models get more complicated, it is useful to have these controls over the display to relieve the screen clutter. Leave the window as shown in Figure 9. The model should appear as in Figure 10.

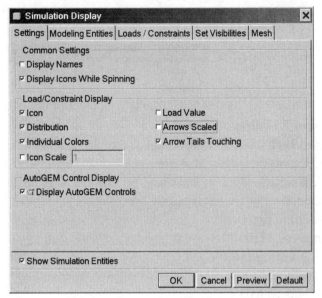

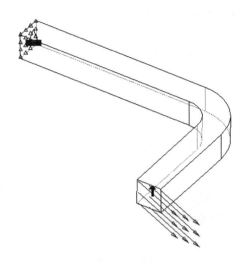

Figure 10 Display with *Tails Touching*

Figure 9 Dialog for setting display options

Specifying the Material

Now we have to tell MECHANICA what the bar is made of. In the right toolbar, select

> *Define Materials*

The Materials dialog window opens (Figure 11). In the list at the left, browse down and click on **STEEL**. Then click the right arrow button in the center. This transfers the material definition to the model, but we have to specifically assign the material to the part. Remember that a model (of an assembly, for example) can have several materials in it, each assigned to one or more different components. Select

> *Assign > Part*

and pick on the bar. It highlights in red. Middle click to get back to the dialog window. If you click the *Edit* button, you can see the numerical values of the material properties. Note that these are in the mm_N_s units system we selected for the part. If you don't like the units in their current format (like tonne/mm^3 for density!), you can select from a pull-down list to get something more familiar (like kg/m^3). *Cancel* this window and select *Close* in the Materials dialog window.

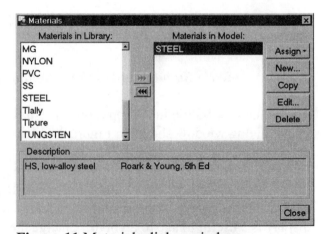

Figure 11 Materials dialog window

Our model definition is now complete. Think back to what we have created so far: geometry, model type, constraints, loads, and material properties. These are the essential elements of the FEA model.

Setting up the Analysis

We now tell MECHANICA what to do with this model. In the top toolbar, select

Run a Design Study

which brings up the **Analyses and Design Studies** dialog window (Figure 12). This window gives you access to all the commands for setting up how you want the model analyzed, how you want convergence to be monitored, where to put the results, and so on. Collectively, these settings define a *study*. The simplest study is a simple static analysis, which is what we will do here. We will get to sensitivity studies and optimization a bit later. This window is also where you actually give the command to run the study - so we are getting close to that!

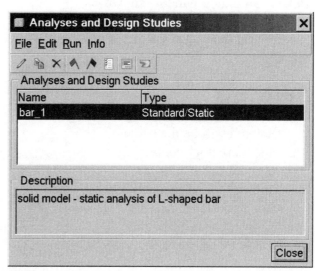

Figure 12 Main dialog window for defining and running analyses and design studies

Most commands and functions in this window have several possible entry points - the pull-down menus at the top or the row of buttons in the middle (see Figure 13). The studies we define will be listed in the central pane just below these buttons. Selecting a study here and using the RMB pop-up gives you yet another way to launch commands.

Obviously, we can have more than one study defined at any given time. Only one of them can actually be running, though.

In the pull-down menu, select

File > New Static

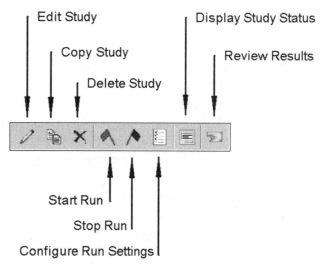

Figure 13 Controls for creating and running analyses and design studies

This opens the **Static Analysis Definition** dialog window shown in Figure 14. Enter a name for the analysis *[bar_1]* at the top of this dialog.

IMPORTANT

The analysis name you just entered will be the name of a subdirectory containing all your result files for this analysis. This name will be important later. It is the *study* name indicated in Table IV of lesson 2.

You can enter a short description for the analysis in the Description box. For a static stress analysis we need to specify/select the constraint and load sets. These are the ones we created above (default names **ConstraintSet1** and **LoadSet1**)

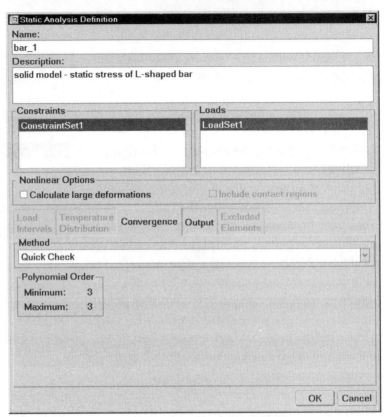

Figure 14 Creating a Static Analysis definition

and are entered automatically. We must also specify what type of convergence analysis we want to use, available in the pull-down list under the ***Convergence*** tab. The types of convergence analysis available here were discussed in Chapter 2. The first run of a model should always be a ***Quick Check***. This does a "complete" analysis of the model at a low (usually 3) edge order. The results of a Quick Check should never be trusted, since no convergence or accuracy information is available. The purpose of the Quick Check is to make sure the model is solvable. For example, if we have improperly constrained the model, the solver will intercept this and give us an appropriate error message. The completed dialog window is shown in Figure 14. Select *OK*. Back in the **Analyses and Design Studies** window, you can see the analysis **bar_1**, type **Standard/Static** as in Figure 12.

If, as in this case, we have only a single analysis defined, we don't need to create a Design Study.

To see if we have any gross modeling errors, in the pull-down menu select

Info > Check Model

This can trap some types of errors in the model (for example, no material specified). We should be informed that there are no errors in the model. If errors have been detected, go back through what we've just done and make sure all the steps were completed. After that, we are on to the next step.

Setting Up and Running the Analysis

This is what we've been waiting for! Still in the **Analyses and Design Studies** window, in the pull-down menu select (or use the toolbar button)

Run > Settings

This opens the window shown in Figure 15. Here is where we can specify where MECHANICA should direct all its temporary and output files. If all goes well, the temporary files are deleted automatically after a successful run. You normally don't bother with them, so sending them to a "trash" directory is all right. Be aware that if this directory is on another computer on a network that you will seriously slow down the operation of MECHANICA . The default output file directory (unless specified in a special configuration file for your system) is the current working directory. Set this to wherever is convenient for you. The dialog window also has an option for specifying how much memory should be allocated to the

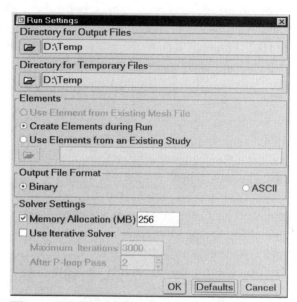

Figure 15 Specifying the run settings

run. In this tutorial, our models will be small enough that the default is satisfactory. If this allocation is too small, MECHANICA will have to page out to the hard disk, which slows it down. A good rule of thumb is to set the **Memory Allocation** to half the physical memory in your computer. Accept the Settings dialog with *OK*.

Back in the **Analyses and Design Studies** window, select (or use the toolbar button)

Run > Start

You are asked if you want error detection. It is a good idea to always accept this, so select *Yes*. The display may flash a couple of times and in the message window you will eventually see[3] "The design study has started." MECHANICA goes through the process of automatically generating the element mesh, formulating the equations, and then solving them. You can follow its progress by selecting the *Display Study Status* button in the toolbar. The **Run Status** window appears with a scroll bar on the right. This is displaying a report file that is stored in the output directory (*bar_1.rpt*). Use the scroll bar to go to the top of the report and browse down through the file. This file is saved after the run.

The automatic mesh generator, AutoGEM, creates 21 solid elements. A lot of other information is also given. Browse through this to see what is there. Near the bottom of the report, in the Measures area, you might note that the maximum displacement magnitude (**max_disp_mag**) is

---

[3] Actually, for this model, by the time you have read this, the run is probably finished!

about 0.5 mm, and the maximum Von Mises stress (**max_stress_vm**) is about 43 MPa. We will compare these results to our "converged" results a bit later. Most importantly, the Quick Check analysis completes with no errors - our model is OK. In the **Run Status** window, select *Close*.

With an error-free Quick Check behind us, we want to change the convergence analysis to Multi-Pass Adaptive. As discussed in the previous chapter, this performs an iterative process. The model is analyzed first using low order elements. An internal algorithm estimates the error in the solution and increases the polynomial order of the offending elements. This process continues until the estimated error is less than the specified tolerance.

Select the analysis **bar_1**, and in the RMB pop-up menu select *Edit*. Now, pick the *Convergence* tab, and in the pull-down Method list, select *Multi-Pass Adaptive*. Set the maximum polynomial order to **9** (the maximum possible), the percent convergence to **5**, and note the radio button beside **Local Displacement, Local Strain Energy and Global RMS Stress**. These settings are shown in Figure 16. We are asking for the convergence iterations (the "passes") to proceed until all of these quantities change by less than 5% between iterations. We will let the element polynomial orders go as high as 9 if necessary (note the default was 6). Accept the settings with *OK*.

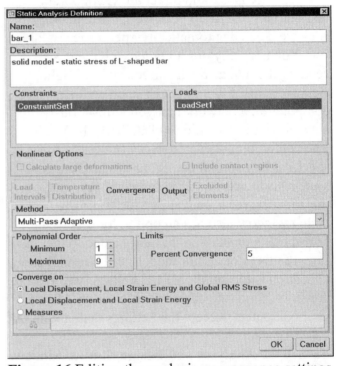

Figure 16 Editing the analysis convergence settings

Now open the *Run Settings* dialog window again. Note the check mark beside "Use elements from an existing study" and the location of the study. We don't need to generate the elements again, since our model has not changed. Close this and select *Start Run*. MECHANICA detects the output files from the previous run (the QuickCheck). Delete these files, accept error detection (probably not needed this time, but does no harm), and the MPA analysis will start up. Open up the *Study Status* window to observe the iterations as they happen. The run will take a few seconds more than last time.

The run converges on pass 8 with a maximum edge order of 8. A great deal of data is presented about the converging solution. We will be plotting some of this in a minute or two. The maximum displacement magnitude is 5.56E-01 (mm) and the maximum Von Mises stress is 4.33E+01 (MPa). Compare these to the results of the Quick Check obtained above. It is very unusual for the results to be so close together. We will see why this occurred a bit later. Before we leave, take note of the elapsed time and CPU time for this run[4] - we will compare these with some runs we'll make a bit later. *Close* the **Run Status** report window, then the **Analyses and Design Studies** window.

Displaying the Results

Creating Result Window Definitions

As mentioned in an earlier lesson, FEA produces an enormous amount of output data. The results directory for the analysis **bar_1** just completed is around 1MB in size for this very simple model. For more complex models, it is not uncommon to have results directories up to a hundred megabytes in size (or more). Reviewing and interpreting this data is best done graphically.

The FEA results can be presented in many ways. Primarily we are interested in various views of the deformation (either displacement components, total displacement magnitude, or material strain components), the stresses in the model (components of normal and/or shear stress, von Mises stress), and data that illustrates the convergence behavior of the analysis. For modal analysis, we can display the mode shapes; we can also produce shear and bending moment diagrams for beam elements. Some of this data represents the primary solution variables (in the case of stress analysis, this is the deformation) while other data (like the material strain and stress) is derived from the primary variables.

In MECHANICA , we display these FEA results in *result windows*. We can display these individually or several simultaneously on the screen. Each result window has a unique name and the contents are determined by a *result window definition*. The set of result window definitions can be saved in an *rwd* file for later use (ie the window definitions don't have to be created again if the analysis is opened up later). Default settings for background colors, text fonts, and so on can also be modified.

Let's see how all this works. This uses some new functions introduced in Release 2001 which make this a lot easier than it used to be. There are even more new "goodies" in Wildfire 2.0. The procedure for displaying the results is as follows:

- create and name the result window(s)
- identify the location (ie the directory) of the result data for each window
- specify the data to be plotted in each window
- select display options for the data
- show the result window(s)
- use optional controls to manipulate the result display in each window

To start this process, in the top toolbar (or in the Design Study dialog window), select

> ***Review Results***

---

[4] Three years ago, on a 300 MHz Pentium with 256M RAM, CPU time was 63 seconds, elapsed time 67 seconds. Now, on an 800 MHz P3 with 1 Gig of RAM, CPU time has been reduced to 9.1 seconds, elapsed time 12.1 seconds. Ain't progress wonderful?

If you are asked to save the current model, select *No,* since that has been done automatically already. A new window (currently untitled) opens up on top of the MECHANICA main window. The new untitled window contains some new pull-down menus and a new toolbar - see Figure 17. The toolbar commands are more-or-less the most common commands needed from the pull-down menus. Most of these are grayed out at this time, but will become active when the first window is defined and displayed.

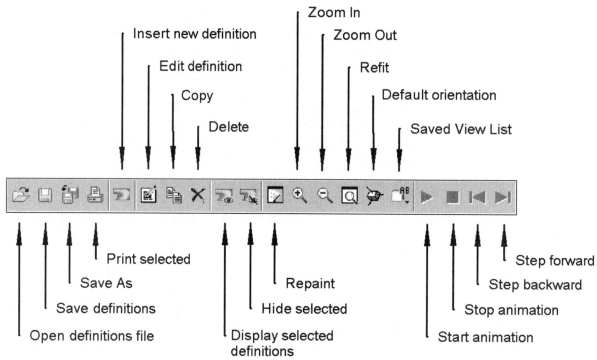

Figure 17 Toolbar for creating and manipulating result windows

The first window we will create contains a color fringe plot of the Von Mises stress. Select the *Insert New Definition* button (fifth from the left in the toolbar), or select in the pull-down menus

Insert > Result Window

In the dialog box that opens (Figure 18), enter a name *[vm]* for this result window. Now we must tell MECHANICA where the data is (the design study or analysis results). You may recall that we named our analysis **bar_1** and in the **Run Settings** dialog we told MECHANICA where to put all the output files. At that location, MECHANICA has created a subdirectory called **bar_1** (ie the same name as the study). Click on the button below **Design Study**, navigate to this directory location and select it, then *Open.* All you need to specify is the top subdirectory named **bar_1** - MECHANICA will find all the files it needs inside there. The dialog window expands for us (Figure 18) to define the contents for the result window.

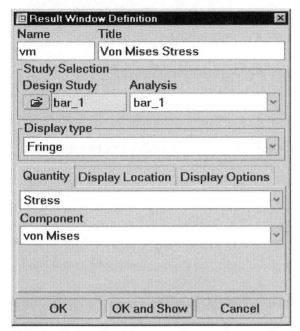

Figure 18 Defining the contents of a result window

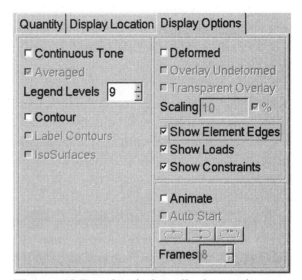

Figure 19 Result window display options (see Figure 18)

In the title field at the top, enter a window title like *[Von Mises Stress]*. The defaults are ***Display Type(Fringe)*** and ***Quantity(Stress)***. The default stress is **Von Mises**. Click on the **Display Options** tab and select the box beside **Show Element Edges**. This will let us see the mesh. Now select the ***OK and Show*** button at the bottom of the dialog window.

The display now shows the fringe plot of the von Mises stress, and most of the remaining toolbar buttons are now live. You can spin/zoom/pan this display using the mouse buttons. You can also manipulate the display using commands in the ***View*** pull-down menu. We will return to this display in a while to explore other options. In particular, the colors might appear a bit murky - we need to fix that (later!).

Now, we want to create a number of other result window definitions. We have two options here:

    ☞ If we use ***Insert > Result Window*** again, we will be asked to specify the results directory again. This is handy if you want to display results from two difference analyses for comparison.

Or

    ☞ If you use ***Edit > Copy*** or pick the ***Copy*** toolbar icon, the result directory and window definition from the current window will be used automatically, saving you several steps and mouse clicks. All you have to do is supply the new window name, and change the desired parts of the definition.

You should experiment with both these options later. For now, since the next window we want will show an animation of the deformation, that is the same results directory, select the ***Copy*** toolbar button. In the new dialog window, enter a name *[def]*. Change the title to *[Deformation]* and set the following options (might as well go for broke here!):

Display Type (Model)
Quantity(Displacement, Magnitude)
Display Options
Shade Surfaces | Deformed | Overlay Undeformed | Transparent Overlay
Scaling 10% | Show Element Edges | Animate | Auto Start
Frames 24

See Figure 21. One item to note here is that the number of animation frames must be a multiple of 4. Too few frames will be jerky, too many will be too slow. You can come back later to experiment with other options in this dialog window. Accept the definition with **OK**. We will show this result window in a minute.

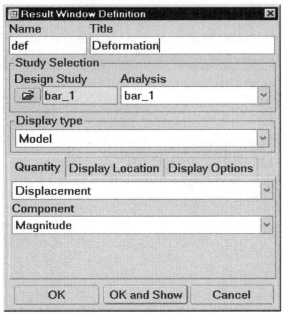

Figure 20 Defining a second result window

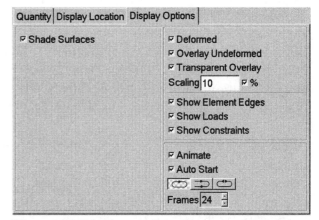

Figure 21 Display options for second result window (see Figure 20)

As promised before, we are going to plot some data to illustrate the convergence process that occurred during the multi-pass adaptive run. We are interested in the stress, deformation, and strain energy. Some key values ("measures") for these quantities are computed and stored for each pass during the run.

Select the *Copy* toolbar button again and enter a name for the new window, *[convm]*. Change the title to *[Von Mises Convergence]* and set the following options. See Figure 22.

> ***Display type(Graph)***
> ***Quantity(Measure)***

Pick the button under the Quantity pull-down list and select

> ***max_stress_vm | OK***
> ***OK and Show***

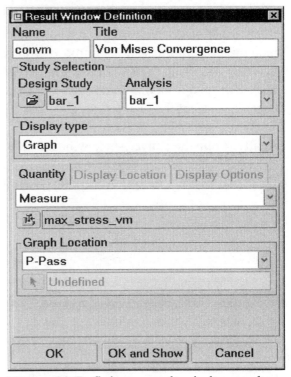

Figure 22 Defining a result window to show convergence of a measure

The screen will now display two result windows side by side (von Mises stress and stress convergence). See Figure 23. The format of these windows has been modified a bit using commands we will see later.

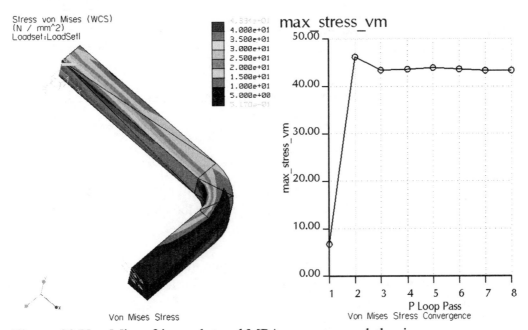

Figure 23 Von Mises fringe plot and MPA convergence behavior

One of these windows is "active", indicated by the yellow border. Click on the convergence graph window so that it is active. Now we can copy this definition to another new window, keeping many of the same definition settings. Select *Copy* to create another window called *[condef]*. The title of this will be *[Deformation Convergence]*. The measure we want to plot is *max_disp_mag* (maximum displacement magnitude) - use the button under the Quantity list box to get it. Accept the dialog with *OK*. See Figure 24.

Finally, *Copy* the convergence window definition to another window called *[constr]*. The title of the window is *[Strain Energy Convergence]*. The measure we want to plot here is at the bottom of the list, *strain_energy*. Accept this definition with *OK*.

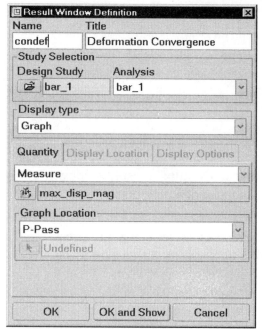

Figure 24 Defining another window to show convergence behavior

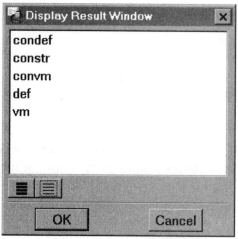

Figure 25 List of defined result windows

We have now defined five result windows, two of which are currently displayed. To see a list of the defined windows, select

>   *View > Display*

or select the *Display Definitions* icon in the toolbar. This will open the window shown in Figure 25. To view the various result windows, pick their name in the list, and then select the *OK* button. Windows can be shown one at a time or several together.

To change any definition settings for a result window, you must first display the window then make it active by clicking on it. To change the window definition, select the *Edit Definition* button in the toolbar.

Let's see what can be done with the individual window displays.

Showing the Result Windows

We'll start with the deformation animation. Highlight only the entry **def** in the list of Figure 25 and then select *OK*. A view of the model appears (wireframe or shaded depending on previous options) animating the deformation. Recall we set up AutoStart in the result definition dialog (Figure 21). All the element edges are shown. Some information is given at the top of the window, and the window title is at the bottom. You can spin/zoom/pan the model using the usual mouse buttons.

Use the animation control buttons in the toolbar to start and stop the animation, and to step forward and backward one frame at a time. Notice that the deformation has been magnified by a large scale factor (listed at the top) for us to see it. Remember that the actual maximum displacement is only about 0.5mm. This is only 1/50th of the bar's cross section width - possibly less than the width of a pixel on the screen. In order to "see" the deformation it must be magnified quite a bit - this is common practice in FEA. The maximum deformation is shown in Figure 26.

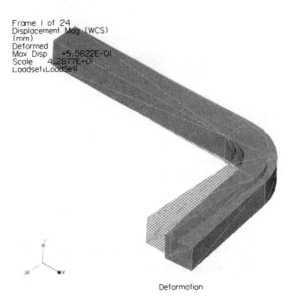

Figure 26 Deformation animation

The information of most importance in an animation like this is whether the deformation is consistent with our applied boundary conditions and loads (or at least what we think they are!). If you make a serious mistake with these (like loads in the wrong direction), it will usually be obvious when you watch this animation. It is useful to produce this animation immediately after running the Quick Check analysis to verify the loads and constraints on the model. Check out the *View > Overlay* and *View > Shade* commands.

Stop the animation, open the **Display Result Window** (Figure 25) and select just the window **vm** in the list, then select *OK*.

A colored fringe plot of the part appears with a legend at the top right. The colors correspond to different levels of the von Mises stress. If your colors appear "murky", select *View* and deselect the *Shade* option. The display is shown in Figure 27. When you are finished exploring this display, let's experiment with some of the options available for viewing the Von Mises stress fringe plot.

Select the *Edit Definition* button, then pick the *Display Options* tab. Change the number of contour levels from 8 to **9**, and select **Continuous Tone**. Turn off the element edges. Accept this window - it should look like Figure 28. This is a very pretty picture (the kind the vendors like to put in glossy sales literature!), but it is difficult to be precise about the color when interpreting the stress levels. This is an example of the kind of post-processing "magic" that can be done. Remember that both Figures 27 and 28 are derived from the same data. Which figure do you think most people would think is "more accurate?" Would you agree?

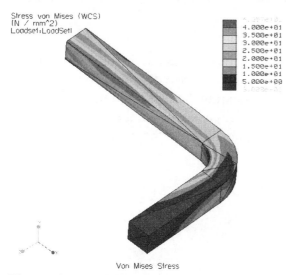

Figure 27 Von Mises stress fringe plot (8 levels, with element edges)

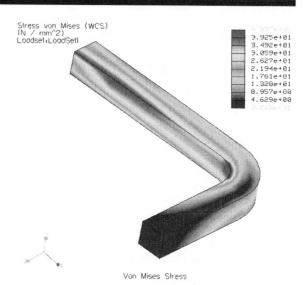

Figure 28 Von Mises stress fringe plot (9 levels, continuous tone, no edges)

In the pull-down menus, select

> *Format > Result Window*

This lets you set the window background colour (white is good for producing hard copy) and the visibility of various elements in the window. Now select

> *Format > Legend*

This shows the minimum and maximum values in the legend scale. This dialog is basically used to set the number of levels (currently 9) and the color spectrum. Go ahead and experiment with these to see their effect on the display. In this dialog you can also set the visibility of the legend, and the minimum and maximum values to be used. The legend values are linearly distributed. When you are finished, return to 9 levels with the spectrum *Structural*.

Edit the window definition for **vm** once again. In the **Display Options** tab, turn on *Deformed* and *Animate* (24 frames). Now *OK and Show* this new definition. You will see a shaded image of the bar, with animated stress fringes. Pretty impressive! Use the animation controls to start/stop and step the animation.

One thing we'd like to know is where does the maximum stress occur? Stop the animation and *Edit* the definition of **vm** again. Leave it as a fringe display, but turn off the deformation and animation. Turn element edge display back on and *Show* the window. In the pull-down menus, select

> *Info > Model Max*

This locates the location (with a small triangle) and gives the value for the maximum stress. You may have to spin the model a bit to see the printing. Now select

Info > Dynamic Query

Put the mouse cursor over the model and move the mouse around. The stress level at the cursor location is shown in the small box to the right of the display. If you click the left mouse button, a marker and label will be put on the display. Beware that if you change the view (by spinning, for example), these markers will disappear. When you are finished, select **Done** in the Query window (or middle click).

In the pull down menu, select

Insert > Cutting/Capping Surfs

We want to cut the model along a horizontal plane halfway between the upper and lower surfaces. This is parallel to the WCS ZX plane. The cutting surface definition is shown in Figure 29. To see the surface, select **Apply**. The model is now sliced on the cutting plane, and we can see the stress levels on that plane, as in Figure 30.

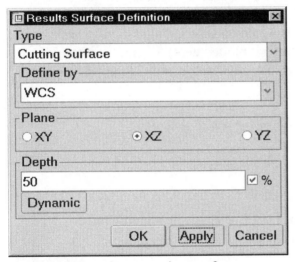

Figure 29 Creating a cutting surface

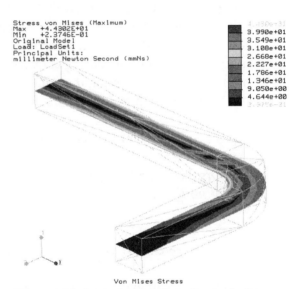

Figure 30 A cutting surface through the model

In the **Results Surface Definition** window, select **Dynamic**. Now click and drag the left mouse button in the display window. You can move the cutting surface up and down in the model. When you are finished, middle click, then **Cancel**. A *Capping Surface* is similar to a cutting surface, except that the material is removed on one side only. Try setting up a capping surface parallel to the YZ plane.

To remove Cutting and Capping surfaces, look for the appropriate **Delete** command in the pull-down **Edit** menu.

Now we will look at the convergence plots. You can highlight all three windows at once in the Display list and then select **OK**. The three graphs are shown in Figures 31, 32, and 33.

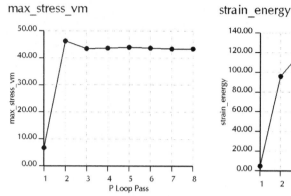

Figure 31 Convergence of Von Mises stress

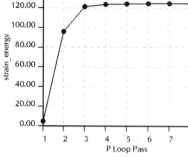

Figure 32 Convergence of total strain energy

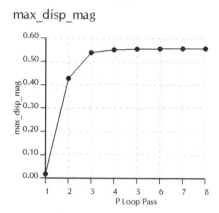

Figure 33 Convergence of maximum displacement

We observe in the convergence plots that, as far as these measures are concerned, the solution essentially was unchanged after the 3rd or 4th pass. This explains why our results are so similar to the Quick Check performed earlier (which used order 3 elements throughout). However, at least one measure being monitored during convergence (but not plotted here) did not meet the 5% criterion that we set. As a result, there must be a least one edge in the model that has been bumped all the way up to 8th order.

To see the p-levels in the model, create a new window called *[plev]* and select

> *Display Type (Fringe)*
> *Quantity (P-level)*

Show this result window and you will see several edges of two elements were 8th order. Curiously, these were not in the region of maximum stress.

Nonetheless, this solution appears to be very well behaved. It is certainly much better behaved than other models you are going to see. Note that we probably could easily limit the maximum edge order to 3 or 4 (thus significantly reducing our solution CPU time) and still obtain very similar results for this model. This is exactly what we will do in the next chapter when performing a sensitivity study and optimization of this bar.

The format of the result windows containing graphs (like those above) can be changed quite easily. Show one of the convergence graphs and then, in the pull-down menus, select

> *Format > Graph*

This opens the **Graph Window Options** dialog window. This dialog has controls for line color and width, text font (style, size, color), axis labels, data symbols, and so on. There is lots to experiment with here. To make any new settings the default for the next time you launch MECHANICA you must do the following:

- ◆ Create an empty text file (before entering MECHANICA) in your start-up directory.
- ◆ In Pro/E, use *Tools > Options* to add the configuration option **bmgr_pref_file**, which should point to the text file created previously.
- ◆ When you are in MECHANICA , set the graph format the way you want then select *Set Default*. All the format information will be placed in the text file and loaded automatically the next time you start Pro/E.

Before leaving the graph window, check out the *File > Export* command. This provides tools for archiving the graph data, or sending it to other programs.

We are finished looking at results for now. Before we leave, however, we should store all our result window definitions so that we don't have to recreate them if we come back to this model later. Use the *Save As* icon on the toolbar of the result window, or use *File > Save As*. This opens a dialog window where you can specify the name and location of the *rwd* file (result window definition) containing all the window settings. A common practice is to name this file the same as the analysis (**bar_1.rwd**) and store it in the same directory as the result directory.

When you have done this, close out the result window and get back to the main MECHANICA window.

Simulation Features in the Model Tree

Open up the model tree. Note that the simulation features (loads and constraints) are all listed. Expand the tree and you will see the members of the load set and constraint set created above. It will appear as in Figure 34. If you right click on these entities and select *Edit Definition*, you can change any of these settings directly out of the model tree. This is a very handy way to make changes to the simulation parameters.

Keep this in mind when you do some of the exercises at the end of this lesson.

Close the model tree.

Figure 34 Model tree with simulation features

Exploring the FEA Mesh and AutoGEM

The FEA mesh created by the automatic mesh generator, AutoGEM, is shown in Figure 35. Let's see how to create this image. Put the display into hidden line mode. Then, in the pull-down menus, select

View > Simulation Display

In the window that appears select the tab

Loads/Constraints

and turn off all visibilities in the **Structure** area (loads, constraints, etc.) - use the button at the bottom to select all. Now use the **Mesh** tab, and select

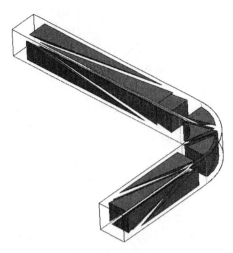

Figure 35 Element mesh created by AutoGEM (default settings)

Shrink Elements 30%

and select *OK*.

Now we need to create the mesh. In the top toolbar, select the Create Mesh button, or in the pull-down menus, select

AutoGEM > Create > Create

An AutoGEM summary appears. Close it. The display changes to show the mesh elements. People who have previously used other FEA packages may be alarmed at the geometry of these elements as follows:

- there are not very many elements involved in the model (only 21 tetrahedral elements)
- there are many long, slender elements (high aspect ratio)
- element corners, even within the same element, can have very different angles
- transitions in element size through the mesh are quite abrupt

In an h-code based method, all these would be signs of a poorly constructed and possibly lethal mesh, and would raise serious concerns about the accuracy of the solution. In this section, we will explore different meshes for this model and compare the performance and results obtained. Hopefully, this will give you a bit more confidence in the operation of the program.

Close the **AutoGEM** window. Do not save the mesh.

Before we proceed, let's review our results from the previous multi-pass analysis. This data is also available in the file **bar_1.rpt** in the output directory. Open the Navigator, pick this file and select *Open* in the RMB pop-up menu. Our main items of interest are the following:

number of tet elements	21
convergence (5%) on pass	8
maximum Von Mises stress	4.33E+01 (43.3 MPa)
maximum displacement	5.56E-01 (0.556 mm)
CPU time	9.1 sec (yours may be different)
elapsed time	12.1 sec (yours may be different)

Close the Navigator and Browser. We are going to change the element mesh and compare results. Some of our control over the mesh is through the settings used in AutoGEM, the mesh generator. We will see other controls later. Select (in the pull-down menu)

AutoGEM > Settings

This brings up the dialog window shown in Figure 36. Notice that the default solid element type is **Tetra** (tetrahedra). Select the **Limits** tab in the middle of the window. This produces the options shown in Figure 37.

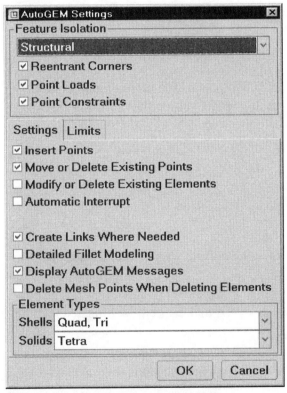

Figure 36 AutoGEM settings dialog window

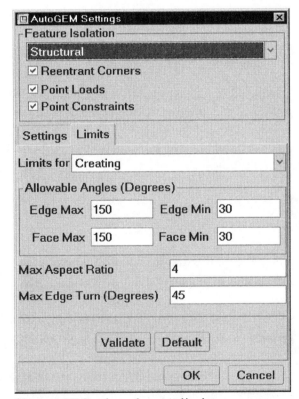

Figure 37 Setting element limits

In the **Limits** area, there are basically three types of settings. The **Allowable Angles** are the angles between edges and faces of an element (see Figure 38). The **Aspect Ratio** is roughly the ratio of length to width of an element. The **Edge Turn** is the amount of arc that can be allowed on an edge (see Figure 39). As you can see, the default settings are pretty broad. This accounts for the wide variation in geometry illustrated in the mesh in Figure 35.

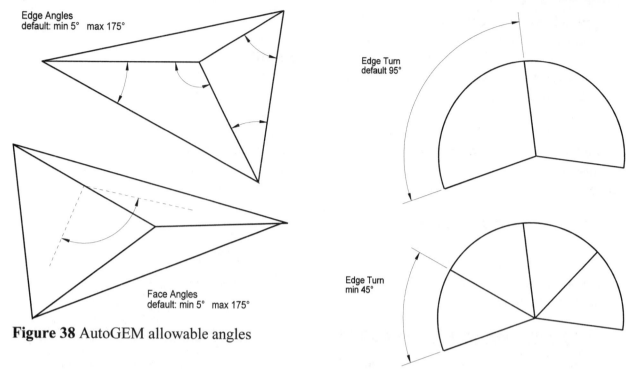

Figure 38 AutoGEM allowable angles

Figure 39 AutoGEM edge turn

If you tighten up on the allowed element limits, you should expect to see more elements in the model. Let's try that. Change the data in the **Limits** dialog to the values shown in Figure 37:

>   Allowable Angles
>   Edge   Max **150**   Min **30**
>   Face   Max **150**   Min **30**
>   Max Allowable Aspect Ratio   **4**
>   Max Allowable Edge Turn   **45**

Then accept the dialog with **OK**. There is no need to save the current mesh since we are about to create a new one anyway.

We will go straight to the MPA (and skip the Quick Check). To make sure of settings, select the **Design Study** button, or in the pull-down menu select

> **Analysis > Mechanica Analysis/Studies**

and **Edit** the analysis **bar_1** to make sure our analysis is the same as before (Multi-Pass, 5% convergence, max order 9). Now select **Run Settings**. Make sure the option to **Use Elements from a Existing Study** is turned off, otherwise the new AutoGEM settings will be ignored and we will use the existing (21 element) mesh. Accept the settings and press **Start** when ready. Delete the existing output files. This run will take a little longer than before - AutoGEM has to work a bit harder this time! Open the **Study Status** window.

The multi-pass run converges on pass 5 with a maximum edge order of 6 with the following results:

number of tet elements	188
maximum Von Mises stress	4.91E+01 (49.1 MPa)
maximum displacement	5.57E-01 (0.557 mm)
CPU time	15.8 sec (yours may be different)
elapsed time	21.4 sec (yours may be different)

Check the **Detailed Summary** box to get even more information about the run. For example, about one third of the CPU seconds were used for mesh generation (see the elapsed time after pass 1).

Use the AutoGEM command to display the mesh as we did before. You can retrieve the existing mesh by using

> *File*
> *Copy Mesh from Study*

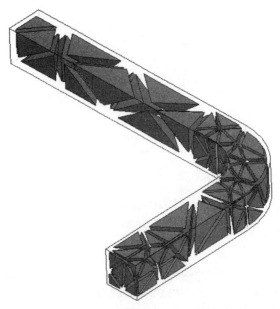

and select the MPA study (top-most directory) we just finished. This saves you some mesh-creation CPU time. The FEA mesh is shown in Figure 40 at the right. There are almost 10 times as many elements in the model as before. As requested, there are no long, slender elements and no very small or very large edge angles.

Note that the run converged two passes sooner with a lower maximum edge order than the previous mesh. Generally speaking, the more elements in the mesh, the lower is the element

Figure 40 Mesh with custom AutoGEM settings

order required for convergence. This is useful to know when you come across a model that will not converge even with the maximum edge order of 9 (NOTE: these cases are very rare!). With lower orders, there are fewer equations per element, but there are more elements. In this case, the net result is that, although there are 10 times as many elements, the computation time is (only) about 2 times what it was before. The relation between mesh size, edge order, and run time is very complicated!

The Von Mises stress and maximum displacement values are within a few percent of the previous results (displacement result is different by 0.001mm!). We could probably say that the results are essentially equivalent. Convergence plots for the run are shown in Figure 41.

Observe in Figure 41 what would happen if this problem was solved using a FEA model having only first order (linear) elements. Many (naive) users might be satisfied with the mesh shown in Figure 40 and probably with these results too. However, without doing a convergence check, they would be stuck with the results obtained on pass 1, not realizing they are in error by almost 50%.

There are clearly two lessons to be learned here. The first is a general observation: FEA results are useless unless accompanied by an examination (and demonstration) of convergence of the solution.

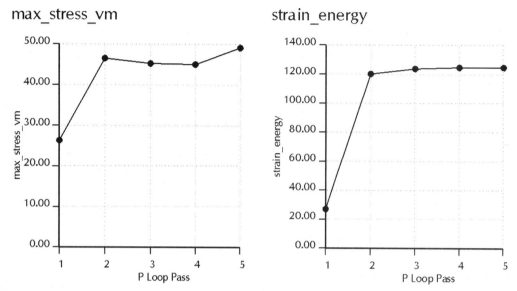

Figure 41 Convergence of Von Mises stress (left) and total strain energy (right) using custom mesh (188 elements)

The second lesson is that, at least in this problem, the mesh density in MECHANICA did not have a large effect on the results in the final solution, although it did significantly affect the solution time. The extreme variations in mesh geometry (aspect ratio, skewness, mesh transition, etc) observed in the default mesh do not cause problems in the solution, as would happen in an h-code solution. In the future, you can probably be comfortable using the default settings for AutoGEM. There may be occasions when you will have a good reason to modify some of these settings, such as when it is difficult or impossible to obtain convergence with the maximum edge order. An example is the allowed edge turn. In a part with filleted corners in a high stress region, you may want to reduce the edge turn to have more elements along the fillet. The same applies if you have holes in a part in a critical region. Before you change AutoGEM settings (that apply everywhere on the model), have a look at the mesh generated by default. There are new routines in AutoGEM that will automatically use a denser mesh in such areas. See the AutoGEM option "Detailed Fillet Modeling" used in the third exercise at the end of the lesson. There are other new mesh controls that we will discover as we go through the lessons.

Running the Model in MECHANICA Independent Mode

As a point of interest, this same model can be created quite easily in independent mode (if you have the software license for that). Figure 42 shows the model composed of 14 brick elements, which were created manually. The Von Mises stress fringe plot is shown in Figure 43.

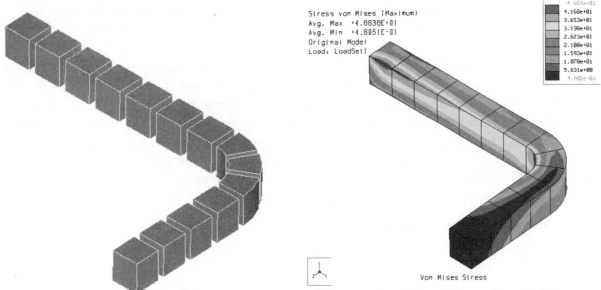

Stress von Mises (Maximum)
Avg. Max +4.6836E+01
Avg. Min +4.8051E-01
Original Model
Load: LoadSet1

Von Mises Stress

Figure 42 Brick elements created in MECHANICA independent mode

Figure 43 Von Mises fringe plot of brick model (independent mode)

The convergence of a multi-pass adaptive analysis on this brick model is shown in Figure 44. The same parameters were used as previously.

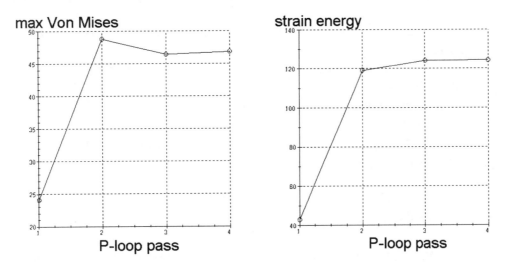

Figure 44 Convergence of model using brick elements in independent mode

The run in independent mode converged on pass 4 with a maximum edge order of 4. Other data is as follows:

number of brick elements	14
maximum Von Mises stress	4.68E+01 (46.8 MPa)
maximum displacement	5.56E-01 (0.556 mm)
CPU time	2.2 sec
elapsed time	2.4 sec

This is two-thirds the number of elements created in our default AutoGEM mesh and less that $1/10^{th}$ the number in our refined mesh. The displacement result is identical, and the Von Mises stress is only slightly different (and well within the convergence tolerance we specified). Observe the execution time - this is $1/5^{th}$ the time required by the program using the default mesh in integrated mode and only about $1/8^{th}$ the time required for the modified mesh. Clearly, there are advantages to be gained by setting up and running in independent mode, if it is available. The disadvantage is that a different user interface and set of geometry creation commands must be learned to do this[5]. The model is also not feature based or parametric, as in Pro/E, and it is considerably more difficult to create both the geometry and the mesh (if you are not a regular user). Leave this to the experts!

Summary

In this lesson, we have performed a complete static stress analysis of a simple part, using solid elements. We have seen all the major steps involved in carrying out the solution and most of the important MECHANICA command menus and dialog windows. To recap, these major steps were the following:

1.   Create the geometry in Pro/E, then launch MECHANICA .
2.   Select the model type.
3.   Complete the model with required data for FEA - constraints, loads, material properties.
4.   Define an analysis and specify mesh and convergence properties.
5.   Run the analysis.
6.   Define and review the desired result windows.

There are many options to explore in this sequence, and many variations of the result windows which you can investigate on your own (perhaps while doing the exercises!).

We observed the effects of changing the settings for the automatic mesh generator. In general, it is not necessary to modify these.

In the next lesson, we will look at the other two types of design studies (sensitivity and optimization), as well as looking at multiple load cases and superposition of solutions. Some important issues in defining loads and constraints will be addressed.

---

[5] See the book **MECHANICA Structure Tutorial - Independent Mode**, also available from Schroff Development Corporation. The independent mode interface has not changed significantly for Wildfire, so the book for Release 2001 should be adequate.

Questions for Review

1.   What form of geometric primitive does AutoGEM use to create solid elements?
2.   Are there any restrictions on the curvature of the edges of any solid element?
3.   Can you get AutoGEM to create brick or wedge elements?
4.   In the AutoGEM settings area, what is meant by edge angle and face angle? What are the default values for their minimum and maximum values?
5.   See if you can find out how the aspect ratio of a solid element is determined.
6.   How do you tell MECHANICA what the part being studied is made of?
7.   Sketch the constraint symbol and identify the six boxes.
8.   The first analysis usually performed is simply to determine if there are any errors in the model. This is called a _____. How is this set up?
9.   What is the purpose of a multi-pass adaptive analysis?
10.   How do you set up a window to view the results of an analysis?
11.   How can you find the location of the maximum stress in the model?
12.   What is the command for a "live" numerical readout of the stress at the cursor location?
13.   What is the difference between a cutting surface and a capping surface?
14.   What is the main benefit of seeing an animated deformation?
15.   What is meant by "convergence"? How is this set up? What are two ways of examining the convergence behavior?
16.   What is the meaning of the following buttons:

   a)   b)   c)   d)

17.   What is our first remedy to a static analysis design study that will not converge in the maximum number of passes?
18.   What material properties must be defined for a static stress analysis?

Exercises

To help save some time with these problems, geometric models can be downloaded from the SDC web site at <**www.schroff1.com**>.

1. Compute the maximum Von Mises stress and deflection in the steel hook shown in the figures below. Model the hook as a sweep cantilevered out from the top plate. (Do you need to model the top plate?) Dimensions are in inches. Examine the convergence behavior of the model.

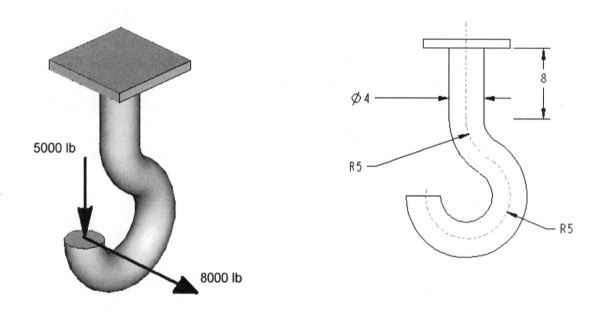

2. Compute the maximum Von Mises stress and deflections in the directions of the loads on the end of the steel connecting rod. The loads are applied at the smaller end. Assume that the inner surface at the larger end is fixed. Examine the convergence behavior. Dimensions are mm.

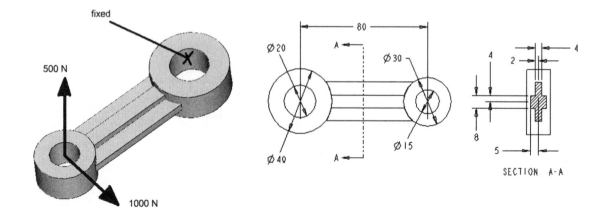

3.   Examine the behavior of the aluminum bracket model below. Units are inches and pounds. The bracket is fixed at each end of the base plate. Look at the stresses, deformation, and convergence. Experiment with the following variations:

   a.   Model the geometry exactly as shown in the figures. Apply a *Uniform | Total* force to the inside surface of the hole using the magnitude and direction given at the right.

   b.   Remove the previous load and try out the *Bearing* option in the Loads menu. This distributes the total load around the bearing surface of the hole.

   c.   Examine the effects of the AutoGEM setting **"Detailed Fillet Modeling"** on the mesh and the results.

   d.   Suppress the fillet and measure the maximum stress and displacement in the model. Plot the convergence graphs (especially the Von Mises stress graph). Comments?

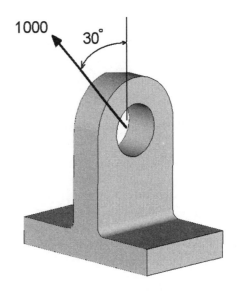

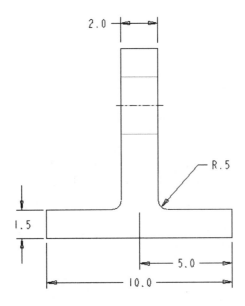

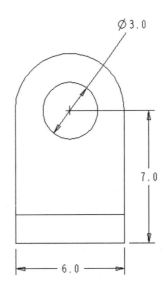

4.   An "interesting" situation occurs if you try to perform an MPA analysis on our L-bracket using the modified mesh (188 elements) and a convergence criterion of 1%, maximum order 9. Try this and plot the usual convergence plots. Where does the maximum stress occur? Can you explain what is happening in the model? (Hint: Poisson's ratio) How could you avoid this problem?

This page left blank.

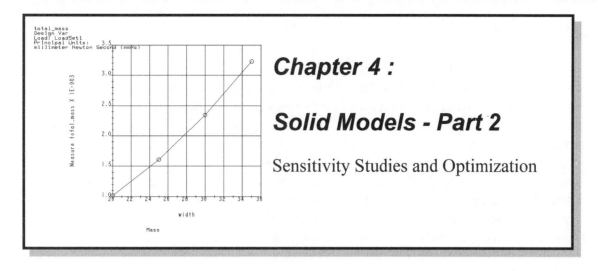

Synopsis

Design variables; sensitivity studies; optimization; considerations for applying loads and constraints; multiple load sets

Overview of this Lesson

In the previous lesson, we used a Standard design study to perform a simple static stress analysis of the part with given loads and constraints. In this lesson, we will first look at the two other types of design studies: *Sensitivity Studies* and *Optimization*. The purpose of these design studies is to automate some of the repetitive work involved in design. This involves specifying one or more design parameters (Pro/E dimensions) that control the geometry of the part. The design parameters can vary over specified ranges. In a sensitivity study, we seek to find out how the variation in a design parameter affects the results of interest (like the maximum stress or deflection). In an optimization, we seek to find the values of the design parameters that will achieve some design goal, like minimizing the part mass, while not exceeding some design constraint, like the maximum allowed stress.

We will also look at some important considerations that must be kept in mind when applying loads and constraints. These are related to the use of p-elements in the solution. Specifically, we'll examine what bad things happen if we specify point and edge loads or constraints in a solid model.

Finally, we will set up and use multiple load sets in order to generalize a solution. This relies on the principle of superposition of solutions.

As usual, there are some Questions for Review and some Exercises at the end of the chapter.

Sensitivity Studies

Suppose you want to find out how a particular dimension (size or location) or model property will affect the results of an analysis. In other words, you want to assess the sensitivity of the model to changes in this parameter. You could do this by manually editing the model (geometry or properties) and performing the analysis many times. The purpose of a sensitivity study is to automate this task.

The general procedure is to set up the model as usual - create the geometry, generate the elements, specify loads, constraints and material properties, and choose an analysis. You then pick the parameters you want to vary and specify the range over which they should change. A sensitivity study is set up, identifying which design variable(s) you want to make active. The study is run and there you have it! Mechanica will automatically increment each specified design variable, manage the model geometry (ie regenerate when needed), and run the designated analysis on the model for each new configuration. You can then set up a results window to show the variation in some measure (like maximum Von Mises stress in the model) as a function of a designated design variable.

Although we used the word "automatic" in the previous paragraph, some subtle problems can sometimes arise in setting up a sensitivity study due to the changing geometry. Chief among these involves the element mesh in the model. You must be cautious about the possibility that certain combinations of design variables may result in impossible meshes. As much as possible, Mechanica tries to use the same mesh throughout the study (to save time). Since, as we have seen, the solution is quite tolerant of large variations in the mesh geometry, this is usually not a problem. However, if the geometric variation is large, we may have trouble. For example, if the design variables are the diameters of two holes in a plate, then the locations of the holes and values of the diameters must not allow the holes to intersect. Also, it may not be possible for AutoGEM to create a mesh (within the current element limit settings) for some combinations of design variables. Mechanica offers tools to check for these types of problems, and the solutions are often easy to obtain.

Launch Pro/E and set the working directory to **CHAP3** (if you have been following instructions to the letter). Bring in the part **bar01** that we used in the previous lesson. In this sensitivity study we are interested in finding out the effect of the bar's cross section dimension on the maximum Von Mises stress and total mass of the part.

Creating a Design Variable

To make things a bit easier in the following, it is useful to change the symbolic names of the dimension parameters in Pro/E. This is done by double-clicking on the part. The dimension for the cross section (25) appears. Select it so that it highlights in red. In the RMB pop-up menu, select *Properties*. Then in the **Dimension Text** tab, change the symbolic name to **width**. Use *Info > Switch Dimensions* to verify the change. Accept this and then change the dimension for the arc on the corner of the trajectory (also currently 25) and to the symbol **bend_radius**. This will be easiest by selecting the curve in the model tree. These modified dimension symbols are shown in Figure 1.

Now we can transfer into MECHANICA:

> ***Applications***
> ***Mechanica***

If you have been playing with the mesh,
go into the ***AutoGEM*** menu and change
the settings back to the defaults.

If you recall in the previous chapter, our
multi-pass analysis **bar_1** converged on
essentially the 3rd or 4th pass. So, let's
modify the analysis so that it stops then. If
your analysis has been deleted, create a
new one called **bar_1** using

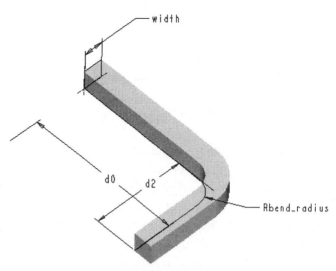

Figure 1 Dimension symbols renamed

> ***Analysis > Mechanica Analyses/Studies > File | New Static*** (or ***Edit***, as required)

Make sure load and constraint sets are selected. Set a **Multi-Pass Adaptive** convergence to
10%. Set the maximum edge order to **5** and ***Accept*** the dialog. We are being a little looser in
our requirements here in order to decrease the execution time for the sensitivity study. Accept
the definition.

Use ***Info > Check Model*** to see if there are any gross errors. You might like to run the new
analysis **bar_1** just to make sure it is doing what we want. The run should converge on pass 4.
The results should be more or less the same as we obtained previously (except that the run will
take a lot less time): maximum Von Mises stress around 4.4E+01 (44 MPa), maximum
displacement magnitude 5.5E-01 (0.55mm).

Creating a Design Parameter

Now we identify a design parameter to vary
during the sensitivity study. In the top pull-
down menu, select

> ***Analysis***
> ***Mechanica Design Controls***
> ***Design Params > Create***
> ***Type(Dimension) > Select***

Read the message window. Click on the part.
Select the width dimension on the cross
section. We return to the **Design Parameter
Definition** window. Enter a description for

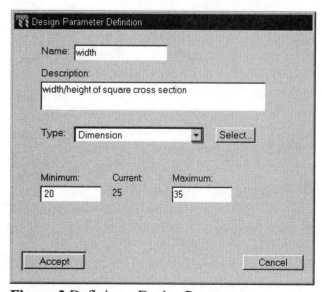

Figure 2 Defining a Design Parameter

the parameter. Near the bottom, enter a minimum value of **20** and a maximum value of **35**. See
Figure 2. ***Accept*** the dialog. In the next window, select ***Done***.

Reviewing the Design Parameters

Will the model "work" throughout the range of the design parameter? In the **DSGN CONTROLS** menu at the right, select

> ***Shape Review***

In the next window, the parameter **width** is listed and a value **20** is shown. Click the ***Review*** button. The part is regenerated with this value - it is hard to see the change in shape here. Restore the part to its original shape and select ***Shape Review*** again. Change the value of the parameter to **35**, and select ***Review*** again. This time, the change in geometry is a bit more obvious. Restore the model to its original shape. We have now made sure that the part can regenerate for the full range of the design parameter. If the parameter value causes a regeneration failure, an error message will inform you of this. Note that this review does not guarantee that the mesh will be suitable for all geometries.

Setting up the Design Study

In the top toolbar, select the ***Design Study*** button then

> ***File > New Design Study***

Call the new design study *[bar_1s]* ("s" for sensitivity). The type is a **Global Sensitivity**. Enter a short description. The analysis to be used is **bar_1**, as defined previously. In the Parameters area, check the box beside the **width** parameter. The default start and end values are Minimum (20) and Maximum (35), set above. Open the pull-down lists to confirm these values. Set the number of intervals in the study to **3** (so that values of **width** = 20, 25, 30, 35 are used). See Figure 3. You can leave the "repeat P-loop convergence" box blank. What this option does (if checked) is force a complete convergence analysis (MPA or SPA) at each new value of the design parameter. This might be necessary if you are concerned that changes in the geometry might cause problems with

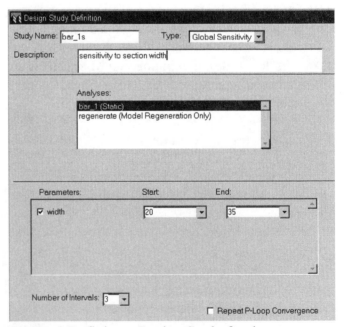

Figure 3 Defining a Design Study for the sensitivity analysis

convergence. If unchecked, the design study will perform a multi-pass analysis for the first value of the design parameter and then use these polynomial orders immediately for each subsequent value of the parameter. This will obviously be quicker. ***Accept*** the dialog.

Running the Design Study

There are now two entries in the run list - our analysis **bar_1** and the design study **bar_1s**. Select the latter, and then *Start Run*. Open the *Study Status* window and watch the proceedings. Observe the convergence analysis is performed only on the first step. For the second and subsequent steps, it goes immediately to order 4.

What is happening to the mesh as the geometry changes? In Mechanica language, we say that the mesh is *associated* with the geometry. This means that the mesh is attached to geometric curves and points, and changes shape as the model changes shape each time it is regenerated with a new parameter value. Mechanica does not have to recreate the mesh with AutoGEM for each new value. This is what gives Mechanica so much flexibility. As we saw in the previous chapter, the p-code mesh is very forgiving of large changes in mesh structure. This will be even more important in the optimization we will perform a bit later.

The run will complete in a couple of minutes. Close the Run Status listing and select *Close* in the **Analyses and Design Studies** window.

Displaying the Results

Select the *Review Results* button to create a couple of result windows. Select the *Insert New Definition* button and call the window *[vm]*. Now, for the Design Study, locate the directory **bar_1s** containing the result files. Fill in the dialog window shown in Figure 4. Mechanica has already guessed that you want to plot a **Graph**. The Quantity is a **Measure**; use the button to pick **max_stress_vm**. The Graph Location is automatically a Design Variable (Mechanica has figured out that **bar_1s** is not a standard analysis), and the **width** parameter is automatically chosen (there is only one in the model). Select *OK and Show* the window.

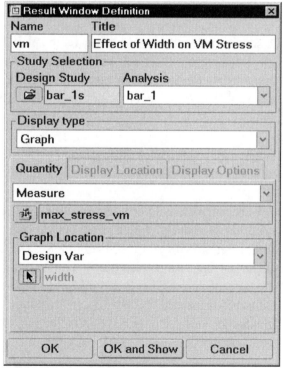

Copy this result window definition to another result window called *[mass]*. Set it up to plot the measure **total_mass** against the **width** parameter. Don't forget to change the title.

Create a third window *[dispY]* that will plot the maximum deflection in the Y direction (**max_disp_y**).

Figure 4 Defining a result window for the sensitivity study

Showing the Result Windows

You can select the three window definitions in the view list and show them all at once. The graphs are shown in Figures 5, 6, and 7.

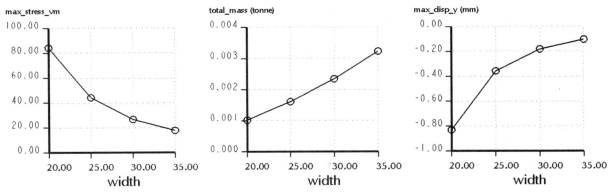

Figure 5 Sensitivity of VM stress to width

Figure 6 Sensitivity of mass to width

Figure 7 Sensitivity of y-displacement to width

As expected, as the width of the section increases, the Von Mises stress goes down, the mass goes up, and the Y displacement gets smaller (notice that the graph is plotting negative values, with positive Y upwards in the model).

A sensitivity study would allow us to select an appropriate dimension to obtain a desired stress level, for example. If several parameters are being investigated, we can see which of them has the largest effect on the stress. To do this, separate studies must be done with each parameter. It is possible to have several parameters included in a single study, but they all are changed simultaneously and the effect of an individual parameter is hard to determine.

When you have one of these graphs displayed, select *File > Export > Graph Report*. This will let you write the graph data to a text file, with extension ".grt".

We are finished with this sensitivity study. Before leaving the topic, recall that we performed a **Global Sensitivity**, which involves varying a parameter over a specified range of values. A **Local Sensitivity** study is used to assess which parameters have the greatest effect on a measure at the current parameter values. This varies the selected parameter(s) by a small amount around the current value. This tells you which parameters will have the larger effects on the particular measure of interest. This is essentially measuring the derivative of the measure with respect to the design variable. For our L-shaped bar, a local sensitivity study with parameters **width** and **bend_radius** shows the following results:

Derivative of ...	(With respect to) width	(With respect to) bend_radius
max_vm	- 4.72	- 0.42
max_disp_Y	0.0534	0.000445
total_mass	0.000133	- 0.0000021

These results show that the width parameter has a significantly larger effect on the model (for these measures, anyway) than the bend radius.

Return to the main MECHANICA window.

Optimization

In this exercise, we want to determine the values of the **width** and **bend_radius** parameters that will result in the minimum weight bar, without exceeding a specified value of stress. From the local sensitivity study we know that:

> increasing **width** results in lower maximum stress but increased mass

and

> increasing **bend_radius** results in lower stress and lower mass

So we have an idea what needs to be done with these parameters. The question is how much to change them to produce an optimal solution. This calls for an optimization study. These are the most computationally intensive studies. Before proceeding with an optimization, you will usually perform a couple of standard analyses to determine the behavior of the solution. For example, you should have some idea as to the convergence behavior of the model. It also helps if you have a pretty good idea of where the optimum solution lies so that you can reduce the size of the optimization search space.

Creating Design Parameters

If you haven't done this already, we need to set up another design parameter for the radius of the bend in the trajectory. In the pull-down menu, select

> *Analysis > Mechanica Design Controls > Design Params*
> *Create > Type(Dimension) | Select*

If you pick on the bar, notice that the width parameter is not visible, since it is already in use. Select the curve that was used for the sweep trajectory (easiest to pick in the model tree). The dimensions will appear. Pick the **bend_radius** dimension. Complete the design parameter definition window, using a brief description. Use a minimum value of **20** and a maximum value of **60**. This will give a fair bit of search latitude. *Accept* the dialog. Then, *Done*.

Examining the Search Space

It is very important to make sure that your part/model will regenerate for all values of the design parameters within the search space. For our simple bar, this should not be a problem. For a more complicated part, changing dimensions of one feature might interfere with the regeneration of another. The more design parameters you use, the more complicated this can get. This is yet another issue that the part designer must keep in mind!

In the last section, we used *Shape Review* to examine the effect of the design variables. Here is something a bit more elaborate. In the **Dsgn Controls** menu, select *Shape Animate*. Check both parameters and set **4** intervals, then click *Animate*. Follow the prompts in the message window. This is not a real (ie moving) animation, but will let you step through the variations in shape between minimum and maximum parameter values (both are changed simultaneously). If you happen to specify parameter values that result in impossible geometries (for example, trying to obtain a bend radius of 150, which is longer than the arm of the bar), the part will not regenerate and you will get an appropriate error message. It is important to do this to ensure that the part is "regenerate-able" over a wide range of geometries, otherwise the optimization routine will stop in midstream. One thing that is not available in integrated mode during this shape review is a look at the elements in the mesh (as you can in independent mode). With very wide variations in parameters, and since the mesh is not recreated for each new geometry but stays associated as discussed above, it is possible that some combinations of parameters may result in part geometries that have illegal meshes. Mechanica will trap this error and force a remesh of the model, which takes additional time of course. The way around this is to not use extremely wide ranges in the parameters or use a Quick Check analysis for the extreme geometries. At the end of the review, restore the model to its original shape.

Creating the Optimization Design Study

Now we need to set up the design study. In the top toolbar, select the *Design Study* button, then

> *File > New Design Study*

Enter the name *[bar_1opt]* and set the Type as *Optimization*. The default goal is to minimize the total mass. Have a look at other possible goals you can set up; for example, you can also maximize a property. In the **Limits on Measures** area, we want to *Create* a limit on the measure **max_stress_vm**. We want this to be less than **35**. Note that our initial design geometry violates this constraint. Check both design parameters **width** and **bend_radius** for use in the optimization. Observe the minimum and maximum values for the search ranges. Change the initial value of both parameters to **25**, the current values. Leave the rest of the settings at their defaults. (You can come back later to see the effects of changing these.) See Figure 8 for the completed dialog. *Accept*.

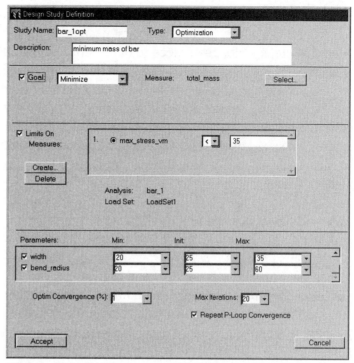

Figure 8 Defining the optimization design study

You can now *Run* the design study, **bar_1opt**. This will take quite a few minutes (maybe 10 or so, depending on your system). You can follow the process in the *Study Status* window. While this is going on, read the next section. The vast majority of the elapsed time for this model is spent loading and unloading Pro/E in the background so that the geometry can be modified. (Look for the trail files in your working directory!) In a more complex FEA problem, the fraction of time spent on these parameter updates would be considerably less.

What Happens During Optimization?

You may be familiar with simple numerical optimization algorithms such as the method of steepest descent. The algorithm in Mechanica is considerably more complex than this, although the basic idea is the same. The algorithm is considerably more efficient than simple steepest descent, and also must contend with the limits (known as *constraints* in optimization theory) in the search space. Mechanica evaluates the current design and tries to decide in what direction to move in the search space (and how far) in order to either remove a constraint violation (like exceeding the allowed stress) or improve on the goal (in our case to reduce the mass).

According to the documentation, you can select from two optimization algorithms: the sequential quadratic programming (SQP) algorithm and the gradient projection (GDP) algorithm. The default is the SQP, which is generally faster for problems with multiple design variables. If the initial design point is feasible (that is, no constraints are violated), the algorithm moves the design point in a direction to better satisfy the goal until/unless a constraint boundary is met in the search space. Then it moves in a direction tangent to the constraint surface, all the while seeking out the minimum value of the objective function. If the initial design point is infeasible (i.e. constraints are violated), then one or more correction steps are taken to reach the (nearest?) constraint boundary. Thus, if the first design is infeasible, the design at the end of these first iteration steps is not guaranteed to be feasible. The GDP has the advantage that, if started with a feasible design, it tends to produce a series of intermediate designs that are always feasible, even if it is unable to locate the global optimum design (either due to the objective function or limits set by you). In contrast, the SQP algorithm does not guarantee that intermediate designs are feasible but only that the optimum (if found) is feasible. The advantage of SQP is its generally increased speed over GDP. For further information on these algorithms, and optimization in general, consult the excellent text <u>Introduction to Optimum Design</u> by J.S. Arora (McGraw-Hill, 1989), Chapter 6.

We will have occasion in the following lessons to experiment more with the settings for an optimization. For now, let's examine what happened with our bar.

Open the ***Study Status*** window and browse through the run report. Notice that the maximum stress constraint is violated for the initial design values. It takes two iterations to find a design that satisfies the stress limit by increasing the section width. The values used for the design variables are given. Then it proceeds to minimize the mass while satisfying the stress limit by increasing the bend radius up to the maximum allowed value. Note the final optimized values for the variables are (width = 27 mm, bend_radius = 60 mm). The mass has actually increased from the initial value, but remember our starting point violated the maximum allowed stress condition. When the run is finished, close out the result window.

Optimization Results

Use ***Results*** as usual to create some result windows. Remember to select the design study **bar_1opt** as the source of the data to be plotted. Set up a window to show the Von Mises stress fringe plot. This will show the stresses in the final optimized design. See Figure 9. Note that the maximum reported stress is 35 MPa as required.

Also, create two windows to plot the measures **max_stress_vm** and **total_mass**. These will automatically be plotted as graphs, with the horizontal axis being the optimization iteration. See Figure 10.

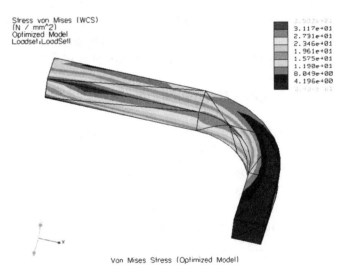

Figure 9 Von Mises stress on optimized geometry

In Figure 10, on the left note the condensed vertical scale. The initial design (pass 0) violates the limit on the Von Mises stress. Two passes were required to reduce the stress to an acceptable level. In order to do that, as seen on the figure at the right, the mass of the part had to increase. As of pass 2, the design was feasible. On the next passes, the optimization was able to reduce the mass somewhat, while maintaining the stress limit.

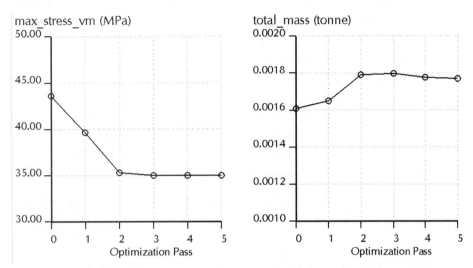

Figure 10 Optimization history (left: Von Mises stress; right: mass)

The values of the design variables during the optimization iterations are shown in Figure 11 (this graph was not generated by Mechanica). This data is stored in the file *bar_1opt.dpi* in the results directory.

We are finished looking at result windows for now. Return to the Mechanica main window.

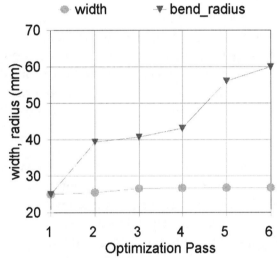

Figure 11 Variation of design parameters during optimization

Viewing the Optimization History

Notice that the display of the part in the Mechanica window is showing the original geometry. We want to transfer the values of the design parameters for the optimized geometry back to the part itself. At the same time, we can view a history of the shape changes during the optimization process given by the parameter changes shown in Figure 11.

In the top pull-down menu select

Analysis > Mechanica Design Controls > Optimize Hist > Search Study

The study **bar_1opt** should be listed. Check it. From here, follow the message prompts. The part is first shown in the original geometry, with the initial values of the design variables (both 25). You can now step through the optimization passes and the part will regenerate with the new design parameter values. Use middle click to advance to the next shape. When you reach the final shape, Mechanica asks if you want to leave the model at the optimum shape. Select *Yes* or middle click. The model now has the optimized design parameters. Leave Mechanica (*Applications > Standard*) and check the dimensions with the *Edit* command. These are shown in Figure 12. We need the original part **bar01** a bit later, so use *File > Save A Copy* and save the optimized part with the new name **bar02**.

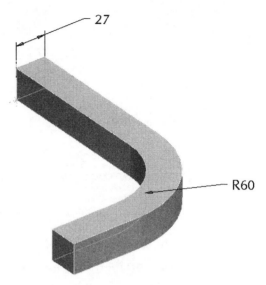

Figure 12 Optimized design

This optimization was pretty simple because the two design parameters didn't affect each other very much. The maximum stress was basically determined by the section width, and the total mass by the bend radius. Our optimized design, in fact, was determined by the maximum allowed bend radius (in optimization terms, the constraint was *active*). We could possibly have guessed that this would be the case, and saved ourselves the trouble of doing the optimization run. However, in other problems, the interaction of the design variables may be much more difficult to predict.

As a final note, in a more complicated design problem, there may be several local optimum solutions surrounding a global optimum. The optimization process used in Mechanica does not guarantee that the global optimum will be found. If this is really what you are after, it is necessary to have a good idea where the global optimum is, and restrict your search (by setting the starting points and ranges of the design variables) to a region close by.

Considerations for Applying Loads and Constraints

You may have wondered why our loads and constraints in the model have been applied on surfaces of the part. Let's do an experiment to see what happens if we deviate from this.

Erase the current model (the optimized **bar01**), and load the original (non-optimum) part **bar01**. (If you don't have the original, just use *Edit* and change the **width** and **bend_radius** values back to 25 and *Regenerate*.) Transfer into Mechanica Structure and open the model tree. Right click on the existing load and constraint sets and *Delete* them.

Set up a new constraint *[fixed_edges]* by fixing all degrees of freedom of only the top and bottom edges on the left end (try this command sequence or use the right toolbar button for an edge constraint):

Insert > Displacement Constraint

In the **References** pull-down list, select *Edge(s)/Curve(s)*, then pick on the upper and lower edges of the face we constrained before (use CTRL-left to pick the second edge), then middle click. Leave everything fully constrained Select *OK*, and you should see the constraint symbols on the edges.

Now create a new load *[point_loads]*. We are going to apply point loads to the top two corners on the other end of the bar (half the load on each). Select (or use the toolbar button for point loads):

Insert > Force/Moment Loads

In the **References** list, select *Points*. We need to create a couple of datum points here. We can do that "on-the-fly" with the menu on the right:

Datum Point Tool

Create two datum points on the end of the bar (see Figure 13), then middle click. Points PNT0 and PNT1 will be displayed (unless that has been toggled off). Complete the Force/Moment dialog window. Enter an X-component of **500** and a Y-component of **-250**. *Accept* the dialog. Note that the load is automatically split between the two points, a consequence of the *Total Load* option that was the default setting under the *Advanced* button.   The model should now look like Figure 13.

Before proceeding, select

Info > Review Total Load

Click on one of the load arrows to identify the load, then select *Compute Load Resultant*. The resultant magnitudes of FX and FY are exactly what we used before. The moments are measured at the specified evaluation point using directions defined by the indicated coordinate system. Close this window.

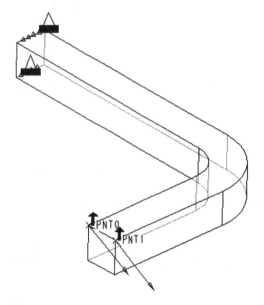

Create a new static analysis called *[bar_1d]* ("d" is for diabolical!). Enter a description and make sure the constraint and load sets are highlighted. Set up a **Quick Check** convergence.

Back in the **Analyses and Design Studies** definition window, delete all studies except **bar_1d**, then select *Info > Check Model*. Read the information window. It looks like we can expect trouble ahead!

Figure 13 Edge constraints and point loads

Now *Run* the analysis **bar_1d**. You will again receive a warning message about the point loads (and they will highlight on the screen). Open the *Study Status* window and see what messages are in there. There is a message near the beginning about "excluded elements", which are discussed below. Other than this, there is not much indication of the serious problem coming soon (for instance, check out the maximum Von Mises stress!).

Now *Edit* the analysis **bar_1d** to run a **Multi-Pass Adaptive** analysis with a maximum order 9 and convergence criterion of 5% on **Local Displacement, Local Strain and Global RMS Stress** as before. *Run* this analysis and look in the *Study Status* window. You will find that the analysis will not converge after all 9 passes. The maximum displacement is a bit larger than we had before, 6.4E-01 (0.64mm). However, the reported maximum Von Mises stress is a whopping 349 MPa (almost 10 times as large as before)! Clearly, there is something peculiar about these results.

To investigate this, create result display windows to show the Von Mises stress fringe plot, and the convergence behavior of the Von Mises stress, maximum displacement, and total strain energy.

Show the Von Mises stress fringe plot (Figure 14). Observe first that the legend is linearly distributed between zero and the maximum stress (in fairly large increments). This reveals the "hot spots" at the locations of the applied loads and constraints, but doesn't tell us much about the rest of the model, which is all the same color. The values of our previous model would have been all included in the lowest couple of fringes. You can select *Format > Legend* to change the values attached to the color fringes. Follow the prompts in the message window to do this. If you have an 8-level legend, set the minimum value to 5 and the maximum value to 40 (see Figure 15). After doing this, you will see that, except for the regions close to the applied loads and constraints, the fringes look pretty much the same as the previous model[1].

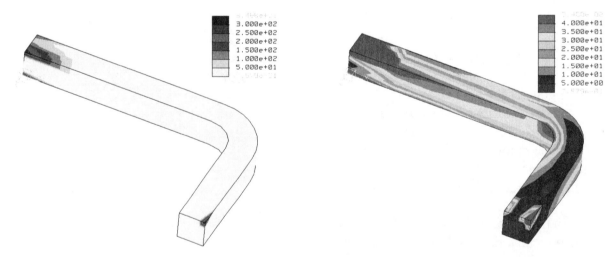

Figure 14 Von Mises stress in the diabolical model

Figure 15 Von Mises stress in the diabolical model (modified legend)

The effect of the applied point loads is quite evident. This is what we were warned about before. Notice the effect of the applied constraints (on the edges at the other end of the bar). These are also causing some stress concentration in the model very close to the constraint. The reported stresses in these areas are very dubious. What is the location of the maximum stress (use *Info > Model Max*)? Is this the same as before?

---

[1] This is an illustration of a rather important principle in strength of materials, with a special name. Can you remember whose name is associated with it?

What is causing this? It is easy to visualize that a force applied on a point (of zero area) will cause an infinite local stress and therefore extremely high gradients of stress as you move away from the point. What Mechanica is trying to do during the multi-pass adaptive analysis is to capture this infinite value and very high gradients. The only way it can do that is to continually increase the element order. If we were to let it go, it would do this indefinitely, since it is impossible to capture this "infinite" value. That is why there is a maximum allowed edge order. Such a point in the solution field is called a *singularity*. The same effect occurs at the location of the constraints in this case. The maximum reported stress in this model is associated with the singularity arising from how the constraint is implemented in the FEA model, not with any real physical behavior.

If you are not particularly interested in what is happening very close to a singularity, these results might be all right. For example, if you are interested only in the stresses in the vicinity of the bend in the bar, then you wouldn't worry.

However, the presence of these numerically induced stress concentrations has seriously disrupted the convergence monitoring within Mechanica[2]. Figure 16 (right) shows the convergence behavior of the maximum displacement. The strain energy graph has the same general shape. This does not show movement towards a constant value. Figure 16 (left), which shows the Von Mises stress convergence behavior, is a clear indication that something is wrong with the model. This is the type of graph you can expect to see if there are singularities in the model - the Von Mises stress will not settle down to a constant value, no matter how many p-loop passes you take, and appears to be increasing indefinitely. This effect is not changed by increasing the mesh density.

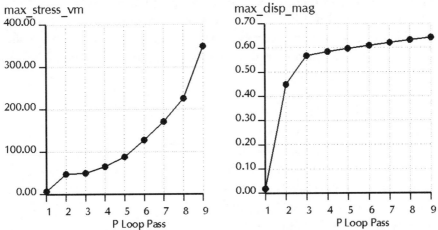

Figure 16 Convergence behavior of displacement and Von Mises stress in diabolical model

---

[2] In independent mode, Mechanica allows you to *exclude* elements near a singularity from consideration in the convergence monitoring. In fact, elements that should be excluded are detected automatically. In this case, stress levels reported at the singularity are ignored.

The lesson to be learned here is that, for solid models, **you should not apply loads or constraints to points or edges - only surfaces**. We will encounter the same type of restriction in other models we will see later. This is a direct result of the p-code method used by Mechanica.

(If you really want to constrain or put a load on a very small area, you can do this by defining a small region on the surface of the model using a datum curve, and applying the load/constraint on the region.)

We do not need to save this model, so you can erase it from the session.

Superposition and Multiple Load Sets

In the problems treated so far, we have provided a single load set to describe the total load on the bar. Suppose that you wanted to analyze the performance of the design under many different loading scenarios. Do you have to analyze each problem separately? This would obviously involve a possibly large number of models and extensive computer time. Fortunately, the answer to this question is "No!" ... read on!

In the solution of our bar problem, the governing equations for the static stress solution are linear, the material properties are assumed to be linear, and the geometry does not undergo a large deformation. Therefore, our entire problem is linear. You probably know that for linear problems, you can make a linear combination of different solutions (this is *superposition*) and the combination will also satisfy the problem (governing equations and boundary conditions). This is a very powerful concept in FEA. In this section we will see how to set up a solution for a multiple of applied loads, and how to superpose the various solutions to analyze an infinite number of loading possibilities. Surprisingly, this does not require a huge increase in computer time.

To get started, launch Pro/E and bring in the original part **bar01** that we used previously. Transfer into Mechanica and delete any loads, constraints, and datum points (if necessary) - do this from the model tree. We previously ran this model with a load FX=500 and FY=-250 applied to the face on the end of the bar. Suppose we wanted to examine the behavior of the bar for many different values and combinations of FX and FY. This is where we can use multiple load sets and superposition.

Constrain the left face of the bar as we did before (all degrees of freedom **FIXED**). Use a surface constraint to avoid the singular behavior we saw in the previous section.

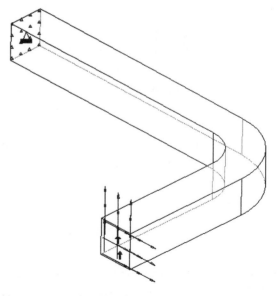

Figure 17 Model with multiple loads

Creating Multiple Load Sets

We are going to separate the load on the end of the bar into components in the X and Y directions, and apply them separately. Start with the ***Force/Moment Load*** button. Beside the **Member of Set** pull-down list, select the *New* button and enter a name *[Xload_set]*, then *OK*. Enter a load name *[Xload]*. Select the button under Surfaces and click on the end face of the bar. Middle click. Enter an X component of **100** and *OK*.

Repeat this procedure to create a new load set called *[Yload_set]* containing a load *[Yload]*. Apply this load to the same surface and enter a value for the Y component of **100**.

It doesn't really matter what values we enter for the components here, since when we combine the loads later we will be multiplying each load by a scaling factor. You could, for example, enter unit loads here. If you do this, the scaling factors become the load magnitudes. You can even reverse the direction of the load by using a negative scaling factor.

The model now should look like Figure 17. Expand the model tree to see the two load sets. You can delete the empty load set LoadSet1.

Setting the Analysis for Multiple Load Sets

We have to set up an analysis definition to tell Mechanica to use both load sets. In the **Analyses and Studies** window select

File > New Static

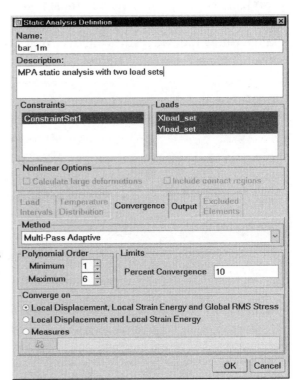

Enter the name of the analysis as *[bar_1m]*. Click on both load sets (**Xload_set** and **Yload_set**) for the analysis. Notice that when you select the second load set, the **Large Deformations** button becomes inactive. (Why?) Set a **Multi-Pass Adaptive** analysis with a maximum polynomial order of 6, with 10% convergence. The definition appears as shown in Figure 18. Accept the definition with *OK*.

In the **Analyses and Design Studies** window, select the analysis **bar_1m**. Check the run settings and, when you are ready, select ***Start Run***. While the study proceeds, open the ***Study Status*** window. Near the bottom of the file, you will see separate results for the two load sets. For load set **Xload_set**, note the maximum displacement is 0.083mm. There is some displacement in the Y direction (1.4E-04mm), which is not expected here due to symmetry of the geometry and the

Figure 18 Creating an analysis with two load sets

load **Xload**.[3] It is possible that our convergence criterion is not tight enough to bring this to zero. For the **Xload_set**, the maximum Von Mises stress is 6.8 MPa. For load set **Yload_set**, the maximum displacement magnitude is 0.14mm and the maximum Von Mises stress is 10 MPa.

Combining Results for Multiple Load Sets

First, we'll create two result windows that will show fringe plots of the stress due to each load set separately.

Create the first window with a name *[vmx]*. Enter the title *[Von Mises - Xload]*, or something similar. To select the design study, find the output directory **bar_1m** and select it. The Result Window Definition changes to show the two load sets. See Figure 19. Select only the **Xload_set** and leave the Scaling set to 1. Keep a fringe plot. In the Display Options, turn on the display of element edges.

Copy this result window to another window, **[vmy]**. This time, select the **Yload_set** only. Once again set up a fringe plot and accept the dialog. Show the two result windows **vmx** and **vmy**. These are shown in Figure 20 below.

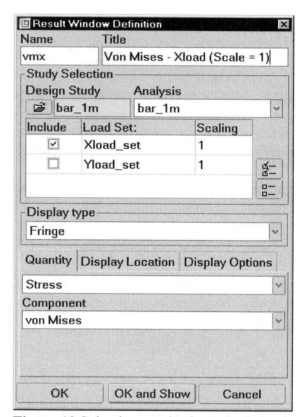

Figure 19 Selecting results for a single load set

---

[3] You might like to create a result window to show the Y displacement, to find out where this value occurs. Is it possibly related to the mesh (which is not symmetric about the mid-horizontal plane)?

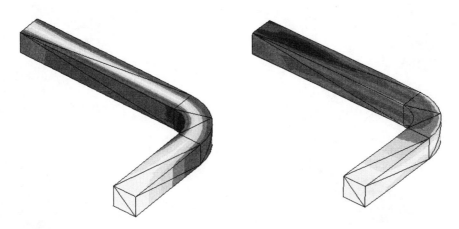

Figure 20 Von Mises stress fringes for Xload (left) and Yload (right)

Now create a new window called *[vmcombined]*. Keep the same design study. Include both load sets. To create the same load conditions that we had previously, the scaling factors should be **5** for **Xload** and **-2.5** for **Yload**. Show the result window.

Figure 21 at the right shows the Von Mises stress fringes for the combined loads. This can be compared with Figure 27 in Lesson 3. Note that the maximum stress is 43.0 MPa, which is slightly lower than before. This is probably due to the change in the convergence criterion and the number of solution passes for the present results.

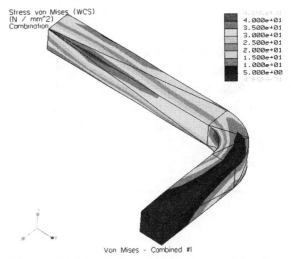

Figure 21 Von Mises stress for combined load (5*Xload, -2.5*Yload)

Copy the definition **vmcombined** to a new window *[vmcomb2]*. Enter a new title ("Von Mises combined #2"). Make the **Scaling** for the Xload **2** and for the Yload **-5**. Accept the dialog and show the window.

Copy the definition **vmcomb2** to a new window *[def2]*. Set this definition up to plot an animation of the deformation.

The windows **vm2** and **def2** are shown in Figures 22 and 23 below. The results are consistent with our expectations.

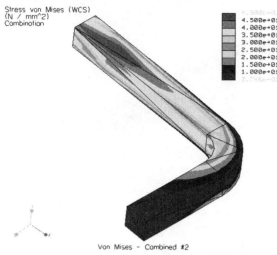

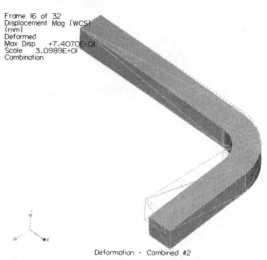

Figure 22 Von Mises stress (combined loads #2)

Figure 23 Deformation under combined loads #2

You see how easy it is to set up the model to handle a multitude of possible loading conditions. Each of our load sets here had only a single load defined in it. It is possible to put several different loads into a single load set (a force and a pressure, for example), but these cannot be separated later. The scaling factor would apply to all loads in the set.

This method of treating multiple loads (with the same constraints) is very efficient, since the problem must only be solved once. Adding load sets does not appreciably increase the solver time. After the solution, various load combinations are put together by post-processing the data, which is a very simple computation. Even if you think loads are related, it does not hurt to put them in separate load sets. This does not cost you much, and gives you a lot of flexibility in result interpretation. The use of multiple load sets is probably one of the most underutilized capabilities in FEA.

Summary

This lesson has covered quite a lot of ground very quickly. We looked at the main steps involved in setting up and running a sensitivity study. The main item of interest here is the creation of design parameters, which is very easy to do. Design parameters are also the key ingredient in performing optimization. For both sensitivity studies and optimization, it is necessary to have a pretty good idea of how the design parameter is used to control the geometry, and the effect it will have on the model. This is particularly true when more than one design parameter may interact, possibly causing an illegal geometry. Optimization can become a very complex problem. Your safest approach is to have some idea of the solution, and keep your search range small. It also helps to keep this process in mind when you are creating the geometry in Pro/E. Try to keep your parts as simple as possible, and you will avoid unpleasant surprises!

We also saw the effects of applying loads and constraints to points and edges of a solid model. The lesson here is - don't do it! Point and edge loads and constraints on solid models will lead to singularities in the solution that will interfere with the normal convergence process (and give erroneous results).

Finally, we looked at the use of multiple load sets. These can save a lot of time if a model must be analyzed under a wide range of applied loads.

This basically concludes our look at solid models. In the next few lessons we are going to look at how idealizations can be used to solve some model geometries much more efficiently.

Questions for Review

1. Give an outline of the necessary steps in performing a sensitivity study.
2. How do you specify a design parameter and its range?
3. What will generally happen to the convergence behavior when you produce a finer mesh in Mechanica? Why?
4. When setting up the design variables, there are two ways that you can determine the geometric effect of changes in a single variable. What are they, and where are these commands available?
5. When running a sensitivity study, what does the "*repeat p-loop convergence*" option do? Under what circumstances would or wouldn't you use this?
6. What is the *shape history*, and how do you obtain it?
7. What should you always do after creating one or more design variables?
8. What search algorithm does Mechanica use in optimization?
9. How can you plot the variation in a measure during the optimization?
10. How do you save the final optimized design?
11. What is the difference between **Shape Review** and **Shape Animate**? What are these used for?
12. What information is required to set up an optimization design study?
13. How do you carry the optimized geometry determined in Mechanica back into Pro/E?
14. What is the potential problem with the use of point and edge loads and constraints?
15. A model can have several load sets. Each load set can have a number of load definitions. How many constraint sets and individual constraints are allowed?
16. What model condition is required in order to use multiple load sets?
17. How many loads can be set up within a single load set?

Exercises

1. Using the bracket model from Exercise #3 in the previous lesson, perform a sensitivity study to examine the Von Mises stress as the radius of the fillet varies from 0.25" up to 1.0".

2. Optimize the bracket from Exercise #3 in the previous lesson to yield a minimum total mass. The design variables are the thickness of the base plate (minimum 0.25") and upright (minimum 0.5"), and the radius of the fillet (minimum 0.25", maximum 1.0"). The Von Mises stress should be limited to 1500 psi.

3. Using the coupler rod (or the hook) from the previous lesson, modify the applied loads to use multiple load sets.

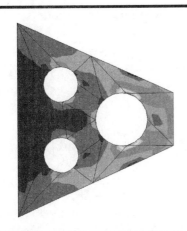

Chapter 5 :

Plane Stress and
Plane Strain Models

Synopsis

Introduction to idealizations; thin plates modeled using plane stress; specifying properties for plane stress models; exploiting symmetry; simulation features; surface regions; coordinate systems; plane strain models; using several materials in the model; applying a temperature load

Overview of this Lesson

As demonstrated in the previous two chapters, the default model type in integrated mode is a 3D solid, for which the finite element model is composed of tetrahedral elements. For many FEA problems, treating them as solids is "overkill." Useful results can often be obtained much more efficiently by treating a simplified version of the problem using *idealizations*. For example, the essence of the problem might be expressed using only a two dimensional geometry. This is the subject of this lesson. Other idealizations (shells, axisymmetric solids and shells, beams) are treated in the next few lessons.

Although Pro/E is by nature a 3D modeler, you can treat 2D FEA problems in integrated mode. These include models for plane stress and plane strain. These idealizations are based on the geometry contained in the 3D Pro/E model. For plane stress and plane strain, as the name implies, the geometry of interest is contained in a planar surface that is "extracted" from the 3D model. The solution is obtained using planar elements (quadrilaterals and triangles) instead of tetrahedral solid elements.

For all problems based on 2D geometry, some restrictions and additional points need to be considered:
1.   All surfaces used to define the geometry for the FEA model must be co-planar.
2.   The geometry must have an associated Cartesian coordinate system. This can be created in MECHANICA.
3.   All entities for the model must be in the XY plane of the coordinate system. This includes all model geometry, constraints, and loads.
4.   For axisymmetric models (these are also 2D models), the model must all be on the positive X side of the coordinate system, X > 0.

In this lesson, the idealizations we will examine are for plane stress and plane strain models. Axisymmetric problems are also based on planar models, but we will defer these to the next lesson. We'll also examine the use of symmetry to further reduce the size of the model (and hence its computational cost). Symmetry is also useful when dealing with 3D problems, but is a bit more complicated. The use of simulation features (coordinate systems, datum curves, regions) is also introduced.

And as usual, there are Questions for Review and some suggested Exercises at the end.

Plane Stress Models

A plane stress problem arises when a model is very thin in one dimension compared to the other two. A typical example is a thin flat plate. Because the part/model must stay in the same (XY) plane, loads must also be applied only in the plane of the plate[1]. In this case, the normal stress σ_{ZZ} (perpendicular to the plate) is assumed to be zero throughout its thickness[2]. Normal stresses σ_{XX} and σ_{YY} are in the plane of the model. The equations that govern plane stress problems are considerably simplified, and plane stress models are among the fastest to compute.

The example problem we will study concerns the stress analysis in the thin flat plate shown in Figure 1. Create this model (call it **psplate**) in Pro/E as follows and according to the dimensions in Figure 2. Use a solid part template and make sure your units are set to mm-N-s. The model consists of a solid protrusion sketched on **FRONT** and some straight holes. Create the lower hole as a mirror copy of the upper hole. To illustrate a point with 2D plane stress modeling here, create the solid plate with a thickness of 20mm. You can delete the default coordinate system.

When the Pro/E part is complete, launch MECHANICA using

>   *Applications > Mechanica*

In the **Model Type** window, select *OK*. This indicates to MECHANICA that we are going to create a default 3D solid model. We will come back in a minute to change this.

Note that the green WCS is created automatically at the default location. As mentioned previously, a 2D model must have an associated Cartesian coordinate system whose XY plane contains the surface geometry for the analysis. We are going to use the front surface of the plate to define our 2D geometry, so we'll need to create a new coordinate system as a *Simulation Feature* there.

---

[1] If there are out-of-plane loads, the model must be treated using shells, which are discussed in Lesson #7.

[2] In fact, all stress components in the Z direction are zero.

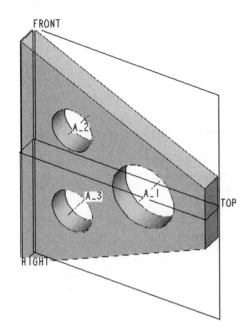

Figure 1 Flat plate solid model

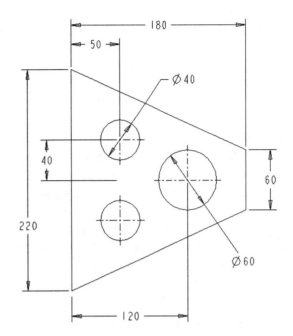

Figure 2 Flat plate dimensions in mm (20 thick)

Creating a Coordinate System

To create the new coordinate system, use the datum toolbar button

> ***Datum Coordinate System***
> ***Type(Cartesian)***

This new dialog window is somewhat different from the coordinate system dialog window in Pro/E. Can you spot the difference?

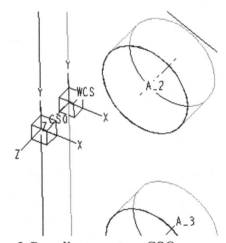

Figure 3 Coordinate system CS0

Holding down the CTRL key, click on the front face of the plate, the **TOP** datum, and the **RIGHT** datum. As they are selected they will be listed in the **Coordinate System** window. A new system will be created at the intersection of these three planes. This is probably not in the desired orientation. If not, pick the ***Orientation*** tab. Select the appropriate surfaces and directions (X, Y, or Z) and use the ***Flip*** button as necessary to obtain the coordinate system **CS0** shown in Figure 3. Open up the model tree to see an entry for the feature there (under **Simulation Features**). When you go back to view the part in Pro/E, this feature will not appear in the model - it is known only to MECHANICA (but will be saved with the model in Pro/E).

Setting the Model Type

Now we can start setting up the plane stress model. In the pull-down menu, select

Edit > Mechanica Model Type > Advanced

This brings up the **Model Type** window that we have seen before (Figure 4). Check the button beside **2D Plane Stress**. This activates the two buttons at the bottom of the window. Select the *Geometry* button and click on the front face of the plate (Note that this is in the XY plane of CS0). It highlights in red. Middle click (or *OK*). Now select the *Coordinate System* button and pick the coordinate system **CS0** we created above. Select *OK*.

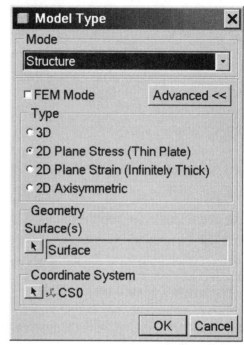

Figure 4 Changing the model type

As you leave the **Model Type** window, a warning window will open up to let you know that if the model type is changed (remember the default was a 3D solid), all previously defined FEA modeling entities will be deleted (loads, constraints, materials, and so on). If that happens, you would have to create them again. Thus, be very careful about selecting the model type, since if you pick the wrong one and have to come back and change it later, much of your work will be lost. Click *Confirm*.

Turn off all the datum axes, planes and so on to remove some of the screen clutter. The surface is now highlighted in purple (this is easier to see in wireframe).

Applying Loads and Constraints

You can now apply loads and constraints in the same way as we did previously. Since we are dealing with a planar model, loads and constraints should be applied to edges (rather than points, which may lead to convergence problems). We will fix the left edge of the geometry and apply a uniformly distributed total load of 100N in the X direction at the right edge. Apply the constraints using the right toolbar button

Displacement Constraint

Call the constraint **[fixed_edge]**. It will be a member of **ConstraintSet1**. Under **References**, select *Edge(s)/Curve(s)*, then pick the left vertical edge of the (purple highlight) modeling surface. Middle click. The only available constraints at the bottom of the window are translation in X and Y for plane stress problems - there is no rotational degree of freedom. Leave both translation directions fixed for this edge. When you select *OK*, the constraint icons will appear along the edge.

Apply the horizontal load using the *Force/Moment Load* toolbar button. Call the load **[endload]** (member of **LoadSet1**). Select the **References(Edge/Curve)** button and pick the right vertical edge. Middle click. Note the default distribution is *Total Load, Uniform*. Enter an X component of **100**. The Z component field is grayed out. Why? Select *OK*. To have the load arrows going outward from the model, select the *Simulation Display* toolbar button and select **Arrow Tails Touching**.

Defining Model Properties

So far we have not specified either the plate material, or its thickness (remember the MECHANICA model is only 2D geometry up to here!). These involve a major variation with what we did in the previous solid model, and with what we will see a bit later with plane strain and shell models. The command picks are non-obvious for this. For 2D plane stress models, we do this using the *New Shell* button

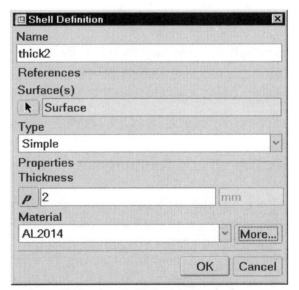

in the right toolbar.

This brings up the menu shown in Figure 5. Enter a shell definition name **[thick2]**. Select the button under Surfaces, pick on the front (purple) plate surface and middle click. Enter a thickness value **2**. The "p" button lets you pick a Pro/E parameter for the thickness (if one has been previously defined). Beside Material,

Figure 5 The **Shell Definition** window to set properties for a plane stress model

select the *More* button, and select **AL2014** from the list and move it over to the right box with the right triple arrow button. Pick *OK*. We don't have to assign the material as we did in the solid model of the previous lesson. The completed **Shell Definition** window is shown in Figure 5. Accept the dialog and open the model tree to see the shell definition entry.

IMPORTANT NOTE:

MECHANICA does *NOT* pick up the plate thickness from the 3D model in Pro/E (unless you use the parameter button). Recall that our actual part has a thickness of 20mm. We will see later that when we use a 3D shell idealization, MECHANICA *will* automatically pick up the thickness from the model. **This does not happen in plane stress!**

Our 2D plane stress model is now complete, and should appear as in Figure 6. Note the additional symbols along the constrained edges. This helps you identify which curves have been constrained.

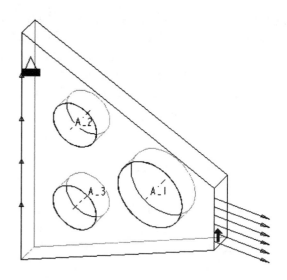

In the top toolbar, select the ***Design Study*** button, then

Info > Check Model

Hopefully, no errors are found. If there are, retrace your steps and resolve the problem.

Figure 6 Complete plane stress model

Setting up and Running the Analysis

We will use the AutoGEM defaults for the mesh. In the **Analyses and Design Studies** window, then, select *File > New Static* and set up a **QuickCheck** analysis called **[pstress1]**. The constraint and load sets should be already selected. Check your ***Run Settings*** (for file directories), and then ***Start Run*** when you are ready. Open the ***Study Status*** window and scan through the run report. The run will not take very long. AutoGEM creates 19 elements (note that they are listed as 2D Plates). No errors are indicated. Make a note of the **max_disp_mag** and **max_stress_vm**.

Assuming no errors were reported with the **QuickCheck**, go back to the *Analyses* menu, and *Edit* the analysis **pstress1**. Change it to a **Multipass Adaptive** analysis with **5%** convergence on **Local Displacement, Local Strain Energy and Global RMS Stress**, maximum edge order **9** (although we hope we don't need that many). Run this analysis. The run should converge on pass 8. The final results should be: maximum displacement magnitude 2.74E-03 (0.00274mm), maximum Von Mises stress 2.84E+00 (2.84MPa). Note the zero displacement in the Z direction and all stress components in the Z direction are zero as well.

Viewing the Results

Create the usual result windows for design study **pstress1**. We are interested in the Von Mises stress fringe plot, animated deformation, and the convergence behavior of the Von Mises stress and the strain energy. When these are defined, display them.

The convergence plots for the multi-pass adaptive (MPA) analysis are shown together in Figure 7. The strain energy rises monotonically to a steady value. The Von Mises stress peaks and then comes down a bit (however note the vertical scale).

A frame from the deformation animation is shown in Figure 8. This also shows the FEA mesh.
Note that AutoGEM has used a mix of triangular and quadrilateral elements here (19 total).
Observe that the left edge is fixed, as desired. What is the deformation scale?

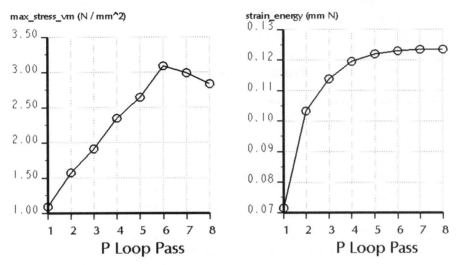

Figure 7 MPA convergence plots: maximum Von Mises stress (left) and
strain energy (right)

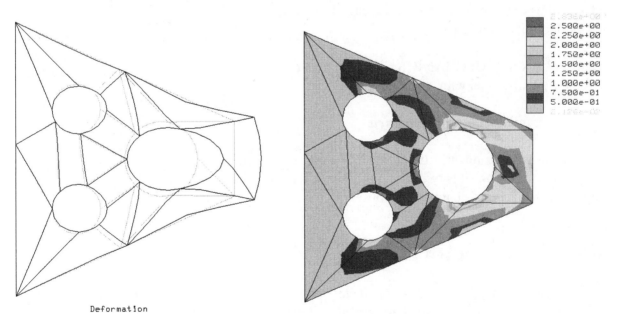

Figure 8 Deformation of the plane stress
model

Figure 9 Von Mises stress

Finally, the Von Mises stress fringe plot is shown in Figure 9. Note the location of the
maximum stress on the large hole and compare this with the deformed shape in Figure 8. Also,
observe that although the mesh is not symmetric, the computed stress distribution is very nearly
perfectly symmetric. The variation is due to two things: 1) we did not go all the way to "perfect"
convergence (try using 1% in the MPA) and 2) the fringe plot is based on values of stress

computed at isolated points (default 4) along each edge[3], then interpolated over the rest of the geometry.

Note that we used constraints and loads applied along edges in this model. We saw previously that for a solid model, surface loads and constraints were fine but point loads and edge constraints lead to difficulties with the Von Mises convergence, related to singularity in the model. For 2D geometries, edge loads and constraints are fine. However, we should anticipate problems if we used point loads and constraints in plane stress.

Exploring Symmetry

In the full plane stress model, we saw that the stress distribution and deformation were symmetric about the horizontal centerline of the model. We can exploit this symmetry in the model to simplify the FEA. In large models, exploiting symmetry can lead to very significant reductions in solution time.

Go back to the Pro/E interface and create a ***Thru All*** cut along the plate centerline from the left edge to the right edge. Remove the material below the cut. The model should look like Figure 10.

Transfer back into MECHANICA Structure with the new symmetric half-model with

Figure 10 Symmetric half-model

Applications > Mechanica

The load and constraint are still defined. The constraint can be left alone, but we will have to modify the load a bit (it is twice too big). We will also create a constraint along the symmetry boundary that is consistent with the observed (or expected) deformation.

Setting Constraints and Loads

Highlight and select the constraint on the left vertical edge. In the RMB pop-up menu, select ***Edit Definition***. This shows the **Constraint** dialog window, in particular, the constraint name **fixed_edge**, and the constraint set **ConstraintSet1**. We are going to add our symmetry constraint into this constraint set. Unlike load sets (where multiple load sets are allowed in a solution), a run can use only a single constraint set, although several can be defined for the model. So, it is important to keep your constraint sets organized and know what you are dealing with. Close this dialog window.

---

[3] This setting is called the **Plotting Grid**, and is set on the OUTPUT tab in the analysis definition window.

For the lower (symmetric) edge select the ***Displacement Constraint*** toolbar button. Name this constraint **[sym_edge]**. It is also a member of **ConstraintSet1**. Set the **References** type to *Edge(s)/Curve(s)* and pick both edges (using CTRL) along the bottom of the model (each side of the hole). Middle click. These are the symmetry edges. The constraint to be applied here is **FREE** in the X direction and **FIXED** in the Y direction. Think about this constraint for a second - what movement does it allow of the model, and is this consistent with our expectations? The model must be allowed to stretch out along X, but we cannot allow a Y deflection since this would imply either opening a split or crack along the symmetry plane, or the model overlapping itself. The constraints along symmetry edges (or planes in a solid model) must be consistent with the behavior of the "missing" part of the model. This can get tricky if the constraint involves rotation! Accept the constraint dialog.

We need to edit the load to the right end. Select the load, then in the RMB pop-up, select ***Edit Definition***. The load is **endload** in **LoadSet1**. It should be a **Total Uniform** force. Change the X component to **50**. We have to divide it in half from the full model. Accept the dialog.

Fortunately, the shell properties (ie the model thickness of 2mm and the material AL2014) we set before are still defined for the model. To confirm that, open the model tree and expand the **Idealizations** entry at the bottom.

The completed symmetric half-model is shown in Figure 11. Note the constraint icon along the symmetric edge. The symbol shows that translation in the X direction is free.

Go to the ***Analyses and Design Studies*** window and create a *New Static* analysis **[pstress_sym]**. In the constraint and load boxes, select **ConstraintSet1** and **LoadSet1**. Set up a **Quick Check**, then *OK*.

Now you can select ***Info > Check Model***. Hopefully, no errors are reported. Otherwise, retrace your steps!

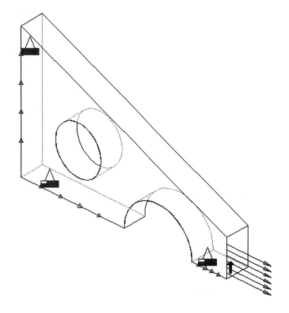

Figure 11 Plane stress model completed

Some readers might note that this model is now over-constrained. How? and Why? A necessary requirement of the constraints is to prevent rigid body motion of the plate. As long as the model is sufficiently constrained against translation (or rotation), it is solvable. Insufficient constraints will cause the run to fail. There is no problem with additional or redundant constraints, as long as they are consistent with each other and our intent. In the present model, with the constraint on Y translation on the symmetry edge, we don't actually need the same constraint on the vertical edge at the left. We have left that constraint in this model so that we can compare results with the full plate. You might like to come back later and FREE the Y translation on the left vertical edge. This would allow a contraction (because of Poisson's ratio) in the vertical height of the plate. On the other hand, keeping the Y constraint on the left edge, and removing the Y constraint on the symmetry edge would allow the model to run, but would not be consistent with our expected deformation. Can you predict what would happen? You might be surprised at the result!

Running the Symmetric Half-Model

Check the *Run Settings* for the analysis **pstress_sym** and then *Start* when ready. Accept error detection. Open the *Study Status* window and look for error messages or other problems. There are only 10 elements in the half-model.

Assuming everything is satisfactory, *Edit* the analysis **pstress_sym**. Change to a **Multi-Pass Adaptive** analysis with the same settings as before. *Run* this analysis, deleting the existing output files. The run will converge on the 8th pass (same as before for the full model), but the run time is significantly less. The results listed near the bottom show a maximum displacement magnitude of 2.74E-03 (0.00274mm) and a maximum Von Mises stress of 2.91E+00 (2.91 MPa). These are essentially the same as before.

Create the result windows for Von Mises stress (fringe), deformation (animation), and convergence of Von Mises stress and strain energy.

The convergence plots are shown in Figure 12. Compare these to Figure 7 for the previous model.

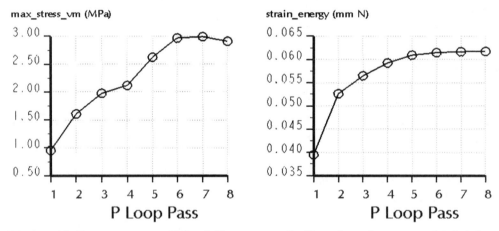

Figure 12 Convergence of Von Mises stress (left) and strain energy (right) for symmetric model

The deformation of the symmetric model is shown in Figure 13. This has the same general shape as the previous model. Observe the deformation along the lower (symmetric) edge. As required, this deflection is only in the horizontal (X) direction. The maximum displacement magnitude is identical to the full model.

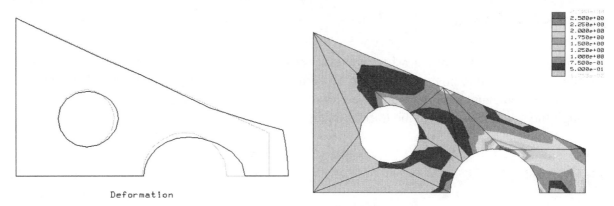

Figure 13 Deformation of symmetric model   **Figure 14** Von Mises stress in symmetric model

The Von Mises stress fringe plot is in Figure 14. Compare this with the upper half of Figure 9. Notice that the fringe legend values have been manually set to have the same values. The results are qualitatively the same. The maximum value reported is somewhat different (about 2%) because of where the analysis stopped with our rather loose convergence criterion. If you rerun both analyses with a tighter criterion (like 1%), the results should be closer together. What problem might occur with the MPA for this model if you use a convergence criterion this small? (Hint: what edge order was required for convergence?) How would you resolve this?

The moral of the story here is that if symmetry exists in the model, then we should exploit it to reduce model size. Symmetry can be used in all model types (solids, shells, beams, and so on). In a complicated model (or in a sensitivity study or optimization) this can cut minutes, even hours, off our execution time. Be aware that symmetry involves all of the following:

♦   symmetric geometry
♦   symmetric loading
♦   symmetric constraints (if not on symmetry plane or edge)
♦   symmetric materials (if different materials exist in the model, the same arrangement must exist on opposite sides of the symmetry plane or edge)

For example, if the **endload** we used on the full plane stress model contained a Y component, then we could not have used symmetry.

In the next exercise, we will exploit symmetry even more - there are two planes of symmetry, and we only need to analyze one-quarter of the full model geometry.

There is one more type of symmetry that will be treated later in these lessons. This involves *cyclic symmetry*, which occurs when a 3D geometry is repeated numerous times in a pattern around a central axis (example: the vanes in a pump impeller).

You can leave the plane stress model and return to Pro/E. Erase the model from the session.

Plane Strain Models

A plane strain problem is one in which the geometry is defined by a 2D shape in the XY plane and the strain normal to this is assumed to be zero. Such a case typically occurs in long thin objects with a constant cross section, fixed total length, and purely transverse loads. Note that although the strain(ε_{ZZ}) is zero, the normal stress in the Z direction is not.

The Model

This example will be used to illustrate a number of new capabilities and functions in MECHANICA that we haven't used before. These include using different materials in the same model (that have very different material properties), applying a temperature load, and using some new simulation features. In order to set up this example, a number of practical (ie. "Real Life") concerns have been neglected[4]. The scenario is the design of a long heat exchanger tube that consists of an inner core like a pipe made of magnesium alloy, and an outer jacket with longitudinal fins made of aluminum. These materials have significantly different elastic properties and coefficients of thermal expansion. The inner pipe is pressurized to 500 psi and the entire model is elevated to a temperature 100°F above the reference temperature. Assume the ends of the tube are fixed so that it cannot expand longitudinally. The geometry of the model is shown in Figures 15 and 16.

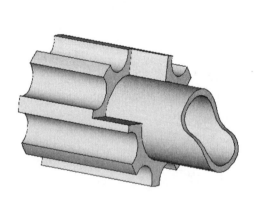

Figure 15 Cutaway shaded view of segment of heat exchanger

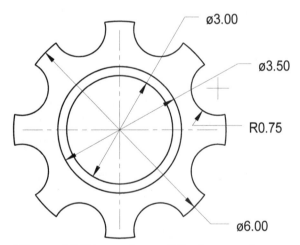

Figure 16 Dimensions of the heat exchanger cross section (inches)

---

[4] In the words of a colleague, this is a solution in search of a problem!

Creating the Pro/E Part

Our first task is to create the model geometry. This will be simplified since we can use symmetry about the horizontal and vertical planes in the cross section. Create a new part file **pstrain1** using the part template **inlbs_part_solid**. The part template has a default coordinate system which we will need. Recall that for 2D models, all the model entities must lie in the XY plane of a reference coordinate system. You can create just the top-right quadrant of this model (as shown in Figure 16) using a single protrusion (blind depth of 5 inches). Use **FRONT** as the sketching plane and extrude the sketch off the back side of the datum (back into the screen). Note that although the inner and outer regions are different materials, we will make a single solid to define the geometry. When we get into MECHANICA, we will extract the 2D shape of the cross section, then create a simulation feature to split the 2D geometry into two regions. Furthermore, the depth of this protrusion is not critical, since all we will be using is the cross sectional shape. See Figure 17 for a view of the solid model.

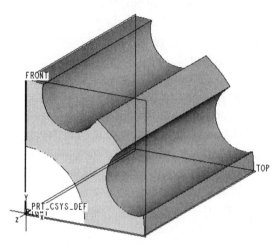

Figure 17 Symmetric quarter-model created in Pro/E

Before you transfer into MECHANICA, use *Edit > Setup > Units* to make sure that your units are set to in-pound-sec (IPS). This is not the Pro/E default (in-lbm-sec). Remember that MECHANICA is very fussy about units!

Now select

       Applications > Mechanica

In the **Model Type** window, select *Advanced* and then *2D Plane Strain*. Pick the button in the **Geometry** area and pick on the front surface of the model (in the XY plane). It highlights in red. Middle click. Now select the **Coordinate System** button and click on the template's default coordinate system. Middle click. Once again you will get a warning window about changing the model type. Select *Confirm*. The front surface should now be highlighted in purple.

Creating Surface Regions

We need to divide the geometry into two regions for the different materials. We do this by defining *surface regions*. One way to create such a region is to use a curve to define a new boundary on an existing surface. We need to make the Ø3.5 curve in Figure 16. There are two ways to do this. We will use a procedure that shows the curve permanently in the model tree. In the right toolbar, select the *Sketch Tool* (one of our regular datum creation buttons). The sketch plane is the front face of the part. Use the **TOP** and **RIGHT** datums for the sketching references. Create a circular arc as shown in Figure 18. When accepted, this will appear as a blue curve on the model. Observe where this is listed in the model tree.

In the right toolbar select the *Surface Region* button, then

Select | Done

You are now in a pre-Wildfire style attributes dialog. Follow the prompts in the message area. First, pick on the datum curve. Then click on the inner portion where the magnesium core will be. The entire surface will highlight in red. Middle click three times to accept all dialogs. The inner surface region highlights in red, and you will find the surface region and datum curve in the model tree. If you click on the model tree entry, the region will highlight on the model. If you pass the mouse cursor over these two regions, they will individually highlight. Notice that we don't have to explicitly define the outer region - MECHANICA will do that automatically.

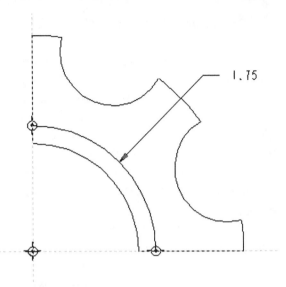

Figure 18 Datum curve

Creating a new Coordinate System

For what we are going to do below with the constraints, it will be helpful to switch from the default Cartesian coordinate system to a cylindrical one. To create a new coordinate system use the datum toolbar button

Datum Coordinate System

In the new window, select *Type(Cylindrical)*. Pick on the existing coordinate system. Accept all the defaults for defining the new system, then *OK*.

The new coordinate system (**CS0**) might be somewhat obscured. Go to the model tree, and select it. Note the shape, color, and name of the new coordinate system icon. The Z axis coincides with the Z axis of the default system. The X-axis label appears as "T = 0" (for θ = 0) and the Y-axis label as "T = 90" (for θ = 90). This is consistent with a right-handed coordinate system. We want to make this our current coordinate system, so select (in the pull-down menu):

Edit > Current Coordinate System

and pick on *CS0*, either on the screen or in the model tree. The cylindrical system now highlights in green to show it is the current one.

Applying the Constraints

We need to apply constraints to the horizontal and
vertical symmetry edges. Think about how these
should be constrained for a moment. The horizontal
edge cannot move vertically, and the vertical edge
cannot move horizontally. Or, combining these two
and using our new cylindrical coordinate system, both
edges can move radially (in the R direction) but not
tangentially (in the θ direction). Since both edges are
constrained in the same way, we can put both edges
into the same constraint. So, in the right toolbar select

Displacement Constraints
References(Edges/Curves)

Use a constraint name *[symedges]*. Select the button
under **References** and pick (using CTRL) on the two
horizontal edges on the lower front of the part, and the
two vertical edges on the left side. The picked edges
turn red. Middle click. Note that the coordinate system
referenced is CS0 (our current system) and the
constraints are in terms of R, Theta, and Z
(MECHANICA knows our current system is
cylindrical). We need to **FREE** the R translation, but
the Theta translation and Z rotation remain **FIXED**.
See Figure 19. Select *OK*. Observe carefully the icon
display for the constraints, indicating that R is free on
both sets of edges.

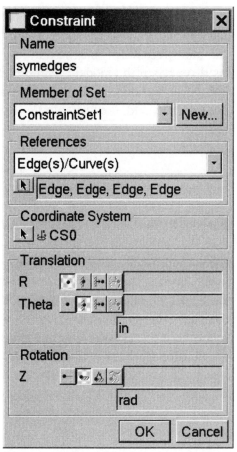

Figure 19 Creating the symmetric
edges constraints

Now we'll set up the loads on the model. So that we can study the pressure and thermal loads
separately, we will use multiple load sets.

Applying a Pressure Load

In the right toolbar select the ***Pressure Load*** button. Select *New* and call the load set **[pressure]**. In the definition window, call the load **[pres500]**. Use the button under Curves and pick on the inside curve of the model[5]. Middle click. Finally, enter a value of **500**. Note that this is a **Uniform** pressure in units of psi (lbf/in^2). Preview the load, then select *OK*. The pressure load will show in yellow.

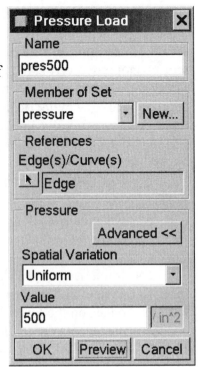

Figure 20 Creating a pressure load

Applying a Temperature Load

For something a little different, we will apply a temperature load to the model. We will apply a uniform ("global") temperature to the entire model. Using MECHANICA Thermal, it is possible to compute a temperature distribution through the model subject to specified thermal constraints, then apply this temperature distribution to the model in Structure. We will do this in a later lesson. By itself, in Structure, the temperature we specify is the difference in global (the entire model) temperature from a reference temperature (nominally 0), for which we assume the model is unstressed[6].

In the right toolbar select

New Global Temperature Load

The temperature load will be contained in a *New* load set called **[thermal]**. Enter a load name **[tempload]**, model temperature of *100*. Since we are in the IPS unit system, this temperature is in °F. This sets the change in temperature from the rest state (reference temperature of 0). See Figure 21. Select *OK*. A small purple symbol appears on the model to indicate that a temperature load has been applied (unless you have turned off the display of load/constraint icons and values).

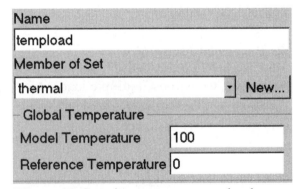

Figure 21 Creating a temperature load

---

[5] If this was a solid model, the default reference type would be a surface.

[6] It is also possible to define material properties that are functions of temperature.

Specifying Materials

Here is where we see how our surface regions are utilized. In the right toolbar, select

Define Materials

Move the materials **AL2014** and **MG** (a magnesium alloy) from the library into the model. Now highlight **AL2014** in the model list on the right and select

Assign > Face/Surface

and pick on the outer region of the model (on the surface outside the datum curve). It highlights in red. Middle click. Now select the **MG** material in the model list, then

Assign > Face/Surface

and pick on the inner region of the surface (inside the datum curve). Middle click, and *Close* the **Materials** window.

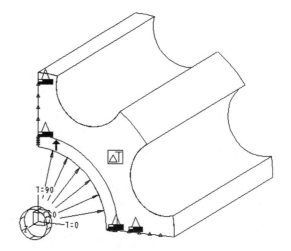

Figure 22 Completed plane strain model

This completes the creation of the model. It should appear as in Figure 22 (some labels have been turned off in the figure). Go to the *Analyses and Design Studies* window and select *Info > Check Model*. If everything has been done properly, there should be no errors at this time.

Running the Model

Quick Check Analysis

As usual, we do a Quick Check to make sure that the stated problem is solvable. Create a *New Static* analysis called **[pstrain1]**. Enter a description. Select the constraint set **ConstraintSet1** and the load sets **pressure** and **thermal**. Select a **QuickCheck** convergence. Now you can select

Run Settings

Check the directories for output and temporary files and RAM allocation. *Accept* the settings and then

Start Run

Accept error detection as usual. Open the *Study Status* window and look for error/warning messages. Note that AutoGEM creates just 5 (!) elements, identified as 2D Solids. There do not appear to be any problems. The results show the maximum Von Mises stress for the pressure and thermal loads are 1,320 psi and 14,000 psi, respectively. Clearly, the thermal load dominates this model.

To make sure the constraints are set up properly, use **Review Results** to create windows showing the deformation animation and the Von Mises stress. Use the combined load sets in the design study **pstrain1**. Include both sets with Scaling set to 1.

Show the deformation animation (Figure 23) and note the agreement with the applied boundary conditions on the horizontal and vertical edge.

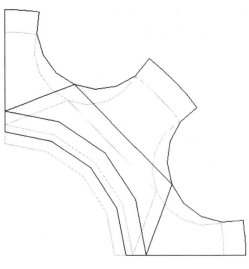

Deformation (combined loads)

In the Von Mises stress fringe plot, we would expect symmetry about a 45° degree line. Note that the display shows line segments around the curved arcs. These connect the plotting grid points that are interpolated along each edge. The actual model does use curved arcs. Close the result windows.

You might be concerned about the low number of elements. To review material from the previous lesson, let's increase the number of elements in the model to see what effect this has. In the pull-down menu, select

Figure 23 Deformation of plane strain model

> **AutoGEM**
> **Settings**
> **Limits**

and change the maximum allowed edge turn to 30. This should cause at least three elements along the interior arc of the model and six along the scallops on the outer edge. Accept the remaining settings.

Edit the analysis **pstrain1.** Select the **Output** tab and change the **Plotting Grid** to 8. Changing this setting should be done with caution since it affects the amount of data that must be generated, stored, and manipulated during post-processing.

Re-run the Quick Check analysis. This time there are 14 elements. No problems are indicated. The maximum Von Mises stresses for the load sets are around 14,100 psi (thermal load) and 1,500 psi (pressure load). These have gone up a bit, as expected with more elements at the same edge order.

Multi-Pass Adaptive Analysis

Assuming there were no errors in the QuickCheck, in the **Analyses and Studies** window, highlight the study **pstrain1**, then in the RMB pop-up menu select **Edit**. Change to a **Multi-Pass Adaptive** analysis with **5%** convergence with a polynomial order maximum **6**. Leave this menu and start the run.

Delete the previous files. Open the ***Study Status*** window. The analysis run converges in 5 passes with a maximum edge order 5. Note the execution time - only a few seconds. The maximum Von Mises stress is 1590 psi for the pressure load and 14,070 psi for the thermal load. These are very clost to the results for the QuickCheck run above. This deserves some thought and investigation of the convergence graphs (left up to you to do!).

Viewing the Results

Since the model mesh has changed, you will have to recreate the result windows to show the Von Mises stress and deformation. Create separate windows for the pressure and temperature loads, and for a combined load with scale factors of 1 on each load set. You can make good use of the ***Copy*** and ***Edit*** commands here when setting up the result windows. For the Von Mises fringe plots, check the **Continuous Tone** option.

First, display the Von Mises stress for the combined loads. See Figure 24. First, notice the changed mesh and affect of the plotting grid. Then, notice that the stress is very uniform in the inner magnesium and also throughout the aluminum although it shows a slight concentration on the innermost part of the fin arcs (where the maximum stress is located). Be careful interpreting these stresses - observe the minimum value indicated on the fringe legend (it is far from 0!).

The deformation for the combined load is shown in Figure 25. This, once again, confirms that our boundary constraints are being obeyed.

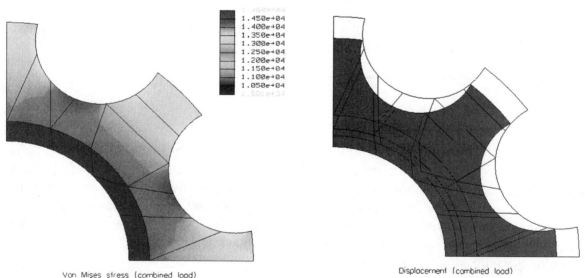

Von Mises stress (combined load)   Displacement (combined load)

Figure 24 Von Mises stress (combined loads)   **Figure 25** Deformation (combined loads)

The Von Mises stresses for the separate load sets are shown in Figure 26. Note that the legend scales are different. The stresses due to the internal pressure are an order of magnitude smaller than those due to the temperature.

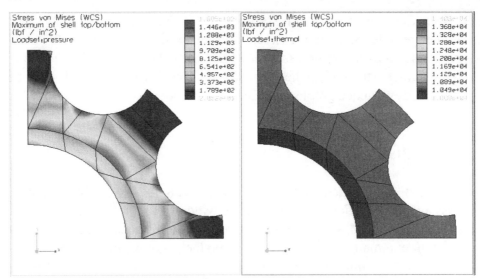

Figure 26 Von Mises stress for pressure load (left) and temperature load(right). Note that the legend scales are quite different.

You might also like to create fringe plots of the normal stress, σ_{ZZ} for both combined and individual loads. This stress is produced since we have essentially fixed the tube so that it cannot expand in length (by assuming plane strain) due to thermal expansion. This puts the tube in axial compression, and is the major contributor to the high Von Mises stress related to the thermal load. Furthermore, the coefficient of thermal expansion of magnesium is greater than aluminum, so much of the load felt by the aluminum is coming from the expanding core of magnesium.

Have a look at the fringe plots for the other normal stresses, σ_{XX} and σ_{YY}. See if you can figure out the physical reason behind these stress values and patterns.

Summary

In this lesson we have introduced two of the possible model idealizations available in MECHANICA: plane stress and plane strain. The main advantage for 2D models, of course, is the speed with which the solution can be obtained versus treating the model as a solid. This is very useful for sensitivity studies and optimization, which are carried out in exactly the same way as previously presented (see the exercises at the end of the lesson). Some restrictions apply for the use of these idealizations, chiefly that the model must lie in the XY (Z = 0) plane of a chosen (or specially created) coordinate system.

We also saw how to set up simulation features (datum curves and surface regions), how to apply different material properties in the model, and how to apply a global temperature load.

In the next lesson, we will look at some more idealizations that utilize a 2D geometry to represent a 3D problem. These are axisymmetric solids and shells.

Questions for Review

1. What are the two element types used in plane stress and plane strain analysis?
2. How do you specify the thickness of a plate for plane stress analysis?
3. How do you assign different material properties to different regions of the same part?
4. For what conditions (what type of model) should "plane strain" be specified/used?
5. For what conditions (what type of model) should "plane stress" be specified/used?
6. What symmetry constraints (if any) would you use for the following models:

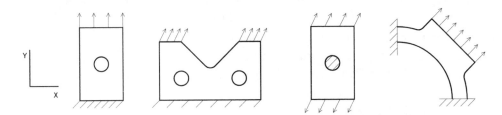

7. What are the advantages of exploiting symmetry in a model?
8. How do you create a coordinate system in MECHANICA? Does this differ from how this is done in Pro/E?
9. Datum points and curves and coordinate systems created in MECHANICA are listed in the model tree under _____.
10. Symmetry in a model includes what entities/features?
11. What are the default element limits for 2D Plates used by AutoGEM?
12. Can a thermal load use a temperature less than the reference temperature?
13. Can a thermal load involve different temperatures in different parts of the model?
14. Can you specify different thicknesses for different regions of a thin plate modeled using plane stress?
15. What happens if you try to specify non-planar surfaces when setting the model type for plane stress or plane strain models? What happens if the surfaces are not co-planar (parallel, but not in the same plane)?
16. Can you apply moment loads in a plane stress model? In a plane strain model?
17. What types of constraints and loads can be expected to cause problems in plane stress models? What is the source of this problem? What is the solution?
18. Does the use of point constraints or loads automatically invalidate results throughout the model?
19. What is the meaning of the following buttons:

a) ![button a] b) ![button b] c) ![button c] d) ![button d]

Exercises

1.   Create a plane stress model of the thin plate shown
     in the figure (dimensions in mm). The plate is 2mm
     thick. This is the "classic" hole-in-a-plate problem.
     Follow the procedures as outlined in this lesson, and
     produce plots of Von Mises stress, deformation, and
     convergence behavior. Set up two models: one of
     the complete plate, and a second that uses symmetry.
     You can compare the results of the two models to
     each other and to analytically obtained solutions, or
     to tabulated values for the stress concentration
     around the hole. What effect does the length of the
     plate have? An analytical solution will likely
     assume an infinitely long plate.

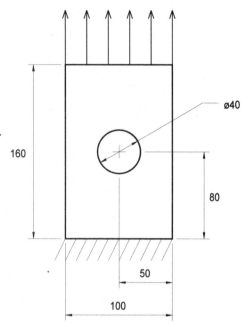

2.   Perform a plane stress FEA of the following geometry and report the stress results,
     maximum deflection, and convergence data. The material is AL2014. Note that there are
     two thickness regions of the thin plate.

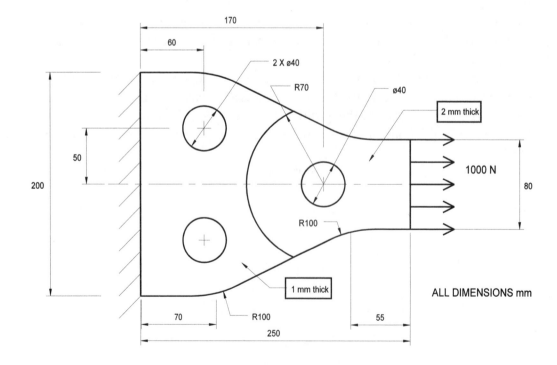

3.   Optimize the geometry of the plane stress model used in this lesson. Use the symmetric half model and select the design parameters shown in the figure below. The objective is to minimize the mass of the plate without exceeding a Von Mises stress of 2.5 MPa. The ranges for the design parameters are

	Min	Max
large hole diameter (**big_diam**)	30	60
small hole diameter (**small_diam**)	30	60
small hole location (**hole_X**)	40	80
small hole location (**hole_Y**)	40	60

Show the optimization iteration convergence plots for the stress and total mass.

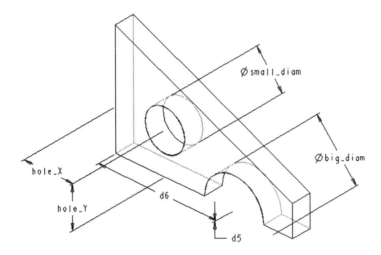

4.   Optimize the geometry of the plane strain model used in this lesson (the heat exchanger). The design parameters are the radius of the datum curve (at the interface between the inner core and outer fins), and the radius of the fin cutout. The objective is to minimize the mass of the model without exceeding a combined Von Mises stress of 14,000 psi. (Do you need to put the pressure and thermal loads into the same load set for optimization?) The ranges for the design parameters are

	Min	Max	Initial
datum curve radius	1.55	1.85	1.75
cutout radius	0.4	1.0	0.75

Show the convergence plots for the stress and total mass. Predict what will happen if you increase the allowable stress to 18,000 psi, then repeat the optimization with that value.

This page left blank.

Chapter 6 :

Axisymmetric Solids and Shells

Synopsis

More geometry commands; axisymmetric models using solid and shell elements; mesh control using detailed fillet modeling, node spacing, and datum curves; pressure loads on axisymmetric models; centrifugal loads; hybrid models

Overview of this Lesson

Axisymmetric models are another type of idealization in FEA. The objective of this chapter is to look at how an axisymmetric model is created using either solid or shell elements[1]. In an axisymmetric model, the three dimensional problem is once again represented using 2D geometry. Some new geometry commands, load types, and MECHANICA utilities will be introduced. A number of ways of controlling the mesh geometry will be explored. We will investigate pressure loads on axisymmetric shells. An underlying message of the lesson is how an FEA model can differ from a CAD model in order to provide a more efficient solution without sacrificing accuracy. We will also look at a model whose geometry is created partly in Pro/E and partly in MECHANICA (a hybrid model).

Axisymmetric Models

Objects with axisymmetry are quite common: pipes, shafts, wheels, tanks, drums, pulleys, and rotational machinery in general. The shape of the object is defined by a planar cross-section (in MECHANICA this must be defined in the XY plane) which is revolved around a central axis (the Y axis). The section does not necessarily need to touch the axis, but cannot cross it. These are sometimes called revolved objects, with which users of 3D CAD programs such as Pro/E will be familiar. Figures 1 and 2 show cutaway views of typical examples of axisymmetric bodies. Axisymmetric models are a close cousin of models with cyclic symmetry. The difference is that the cyclic models consist of a 3D shape that appears in a regular pattern around an axis. A true

---

[1] Mixed models containing both axisymmetric shells and solids are also possible, but we will not have time to look at them here.

axisymmetric model is defined by a 2D shape (area or curve) that is revolved to form a continuous rather than periodic geometry. We will have a look at cyclic models in a later lesson.

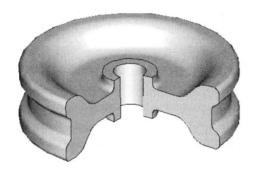

Figure 1 A solid flywheel that would be modeled with 2D Solid elements

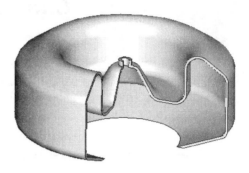

Figure 2 A hollow tank that would be modeled with 2D Shell elements

Elements

Two types of elements can be used with axisymmetric models: *2D Solid* or *2D Shell*. Both element types are defined on a single plane surface that contains the axis of symmetry. The difference between objects for these element types is illustrated in the figures above. The figure on the left would use solid elements defined on the solid cross-sectional area. The figure on the right would use shell elements that follow the cross sectional shape of the thin wall. These could be defined using a single datum curve or the edge (inner or outer) of the solid wall. For both model types, only the two dimensional shape of the cross section needs to be defined. For the axisymmetric shell, you do not even have to make a solid part - just datum curves.

Loads

A number of different loads can be applied to axisymmetric models. The major ones are:

- **Total Load** - If applied on a curve, edge, or shell represents the total load acting on the revolved surface obtained with the entity. If applied on a 2D Solid, represents the total load acting on the revolved volume. In either case, the total load stays the same when the geometry changes during sensitivity or optimization studies.
- **Force per Unit Area or Volume** - If applied on a curve, edge, or shell represents the load per unit area acting on the revolved surface obtained with the entity. If applied on a 2D Solid, represents the load per unit volume acting on the revolved volume. In either case, the total load will vary if the geometry changes during sensitivity or optimization studies.
- **Centrifugal** - Load developed by a rotation around the axis of the object, specified by the rotational speed in radians/sec, and depends on the material mass density.
- **Pressure** - when applied to an edge, creates a force per unit area normal to the revolved surface.

Note the significant departure in MECHANICA from some other FEA packages, where axisymmetric loads are sometimes defined on a *per radian* of revolution basis.

Constraints

In an axisymmetric model, the axis of symmetry is fixed and is always the Y axis of the world coordinate system. This automatically creates a constraint against rigid body motion in the radial direction. The only other rigid body degree of freedom which must be constrained is translation in the Y direction. This will require an explicit constraint in the model. It is also often possible, and desirable if the geometry allows, to use symmetry about the radial (or X) axis.

Restrictions

When setting up an axisymmetric model, some restrictions apply. Foremost among these is that the axis of symmetry must be the Y-axis of the world coordinate system, and all model elements must lie in the right half plane X ≥ 0.

Axisymmetric Solids

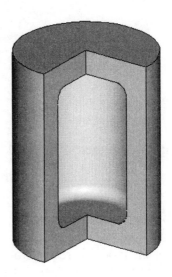

Our first axisymmetric model[2] is the thick-walled steel pressure vessel shown in cutaway in Figure 3. The inside of the tank will be pressurized. Note that the upper and lower inside corners of the tank have a fillet. It happens that the maximum Von Mises stress that occurs in this tank is at these filleted corners. It will be left as an exercise to perform a sensitivity study to examine this. We will model the cross section using 2D Solid elements. At the end of this section we will investigate some methods to control the mesh created with AutoGEM. We will also perform an analysis using 3D solid elements (that is, the MECHANICA default) in order to compare results and solution performance.

Figure 3 Thick-walled pressure vessel

Creating the Model

We can make good use of symmetry in this problem. Since we intend later to do a 3D analysis, we will set up the solid model using all three planes of symmetry (two vertical and one horizontal). In the last lesson in the book, we will revisit this model, treating it as a solid using cyclic symmetry constraints.

The part we are going to create is a 1/8 solid model of the tank that utilizes symmetry about the horizontal midplane, and two vertical planes through the axis of revolution. See Figure 4. Start a new part in Pro/E called **[axitank]**. Change your part units to **in-pound-s** (IPS) using the *Edit >*

---

[2] Adapted from <u>The Finite Element Method in Mechanical Design</u>, Charles E. Knight, PWS-Kent, 1993, pp.165-168.

Setup > *Units* command. Create the base feature as a 90° revolved protrusion off the **FRONT** datum plane. Set the revolve direction so that the protrusion comes off the back side of the datum. We want to use the face of the model lying in the **FRONT** datum (and therefore in the default XY plane) to define our axisymmetric solid geometry. The geometry of the sketch for the revolved protrusion is shown in Figure 5. When the revolved protrusion is finished, create a 0.5" radius round on the inner corner.

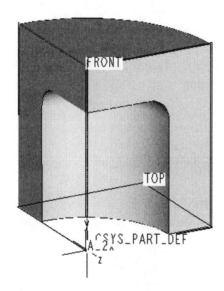

Figure 4 Symmetric 1/8 solid model

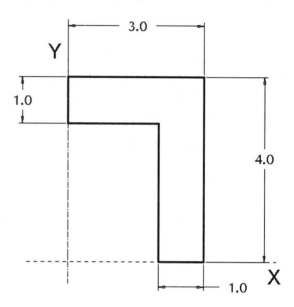

Figure 5 Sketch for the axisymmetric geometry

Setting the Model Type

When the model is ready, transfer into MECHANICA with

> *Applications > Mechanica*

In the **Model Type** window, select *Advanced*, then the *2D Axisymmetric* option. For the Geometry reference, pick the front surface of the part. Middle click. For the Coordinate System, pick the part's default system, then *OK*. Turn off the datum planes.

The surface is now highlighted in purple - this is the geometry for the 2D model.

Applying Constraints

Only one constraint is required, to prevent rigid body translation in the Y direction. We apply this on the lower edge of the front surface (on the X-axis). Select the toolbar button:

New Displacement Constraint

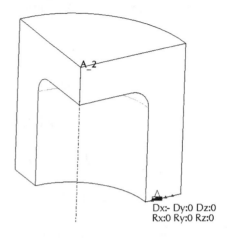

Name the constraint [**midedge**]. It will be in
ConstraintSet1 by default. Set the Reference type to
Edge/Curve and select the bottom edge along the X-axis.
Set the X translation constraint to **FREE** and leave the
others (Y translation, Z rotation) as **FIXED**. This allows
the tank to expand radially but fixes it against rigid body
motion in the Y direction. Why is it necessary to prevent
rotation around the Z direction? Come back later (Exercise
#1) to make this FREE and see if what happens is what
you expect. Accept the constraint with *OK*. The constraint
symbol appears on the model as in Figure 6.

Figure 6 Constrained edge

Applying Loads

Use the *Pressure Load* toolbar button. Create a new load set called *[pressure]*, and name this
first individual load *[p1000]*. Select the button under **References** and pick (using CTRL) on the
three inside edges (including the fillet) of the tank. Middle click. Enter the value **1000**. Note
that the default under *Advanced* is *Uniform*. We could create a hydrostatic pressure field, for
example, where the pressure was a function of postion - useful for very tall tanks holding liquids.
The completed dialog box is shown in Figure 7. Select *OK*. In the model tree, if it is listed there
you can delete the empty LoadSet1.

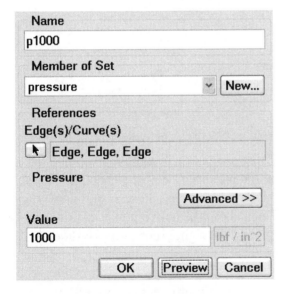

Figure 7 Creating a pressure load

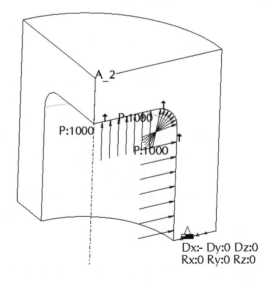

Figure 8 Model completed

Defining Material Properties

In the right toolbar menu select

Define Materials

and move the material **STEEL** to the model list. Then select

Assign > Face/Surface

and pick on the model surface. Middle click. Then *Close*.

Our model is now complete (see Figure 8).

Setting up and Running the Analysis

We should do a Quick Check before proceeding to see if any glaring errors are present in the model.

Design Study > File > New Static

Call the analysis name *[axitank]*. Enter a description. The constraint set is **ConstraintSet1** and load set is **pressure**. For Convergence, specify a *Quick Check*. We are ready to run the analysis, so navigate to:

Run Settings

Select the directories for output and temporary files, then

Start Run > Error detection(Yes)

Click on the *Study Status* button and review the data. There should be no error messages. AutoGEM creates just 3 (!) elements (two quads and a triangle). Note that these are called *2D Solids*. The maximum Von Mises stress is 3876 psi, and the maximum deflections in the X and Y directions are 0.000181 in and 0.000217 in, respectively. Remember that these values don't mean much with the Quick Check analysis since we have no idea about the convergence of this data. It doesn't hurt to have a look at them, since they will be at least the right order of magnitude. A serious blunder might show up in an unusual value here.

Since there are no errors, we can change the analysis method to a multi-pass adaptive setting and rerun the study. Pick the analysis and using the RMB pop-up menu, select

Edit

Change the convergence to **Multi-Pass Adaptive** and set a convergence of 5% on **Local Displacement & Local Strain Energy & Global RMS Stress** with a maximum polynomial order of **9**. Then select

> *OK > Start Run*

Delete the previous output files and, just to be sure, accept error detection. Select the ***Study Status*** button and review the run data. The run has converged on pass 8 with a maximum edge order of 8. The maximum Von Mises stress has increased to 5075 psi, and the X and Y maximum deflections have changed to 0.000184 and 0.000264 inches, respectively[3]. You might note the total CPU time required, since an interesting result will occur shortly.

Viewing the Results

Let's have a look at some results of the run. Set up a window for the Von Mises stress (fringe plot). Under the Display Options tab, select the button to show element edges and **Continuous Tone**. Once the first result window has been set up, use *OK and Show*. Now select the *Copy* toolbar icon to create a new window *[def]* and change window parameters to produce the deformation animation (displacement magnitude, animation, 12 frames, Reverse). The two windows should look like Figures 9 and 10. You can control the visibility of display items in the result windows using ***Format > Result Window*** in the pull-down menu at the top.

Examine these results carefully. Things you should look for are: is the deformation what you would expect? Are the constrained edges behaving properly? Note that the edge on the vertical Y axis stays on the axis even though there is no explicit constraint there. Why? Do stress concentrations occur in the expected location(s)? You might like to set up result windows for the normal stresses in the X (radial), Y (axial), and Z (hoop) directions. On appropriate edges, the XX and YY normal stresses should show values of -1000 psi on the interior surface where pressure is applied, and 0 on the outer surface. The hoop stress on the interior surface can be computed theoretically if end effects are ignored, yielding a value of 2600 psi. In the result window showing σ_{ZZ} what value do you get on the inside surface of the tank at the midplane of the model (on the X axis)? You can use a ***Dynamic Query*** in the fringe plots to review these values.

Before you leave the results area, save the current result window definitions by selecting ***File > Save As*** and specify the full name **axitank.rwd**. We can use these definitions later to save some time as we explore the model mesh.

---

[3] The results obtained by Knight using a first-order h-code analysis with 70 quad elements are: maximum Von Mises = 4021 psi, max deflection X = 0.000177, max deflection Y = 0.000255

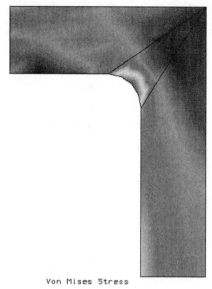

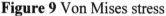

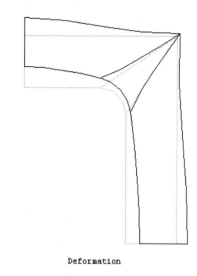

Figure 9 Von Mises stress   **Figure 10** Deformation

Exploring the Model

Changing the Mesh with AutoGEM

You might be concerned that this solution used only 3 elements. Let's investigate the effect of the mesh. Go to the pull-down menu and select

AutoGEM > Settings > Limits

Change the limits for element creation to the following:

Edge Angle:	Max	150	Min	30
Aspect Ratio	Max	4		
Edge Turn	Max	30		

This should increase the number of elements in the mesh, since there is less flexibility available to AutoGEM. Accept these values. Go to the **Analyses & Design Studies** window and in the *Run Settings* dialog, make sure the option "Create elements during run" is turned on. *Start* the analysis. Open the *Study Status* window.

With the new AutoGEM settings, we see that the model contains 33 elements. The run converges on pass 3 with a maximum edge order of 3. The maximum X deflection is 1.83E-04 and the maximum Y deflection is 2.62E-04. These are within a few percent of the previous run. The maximum Von Mises stress is now 5045 psi. This has decreased sightly from the previous run (probably due to where the convergence stopped), but is certainly in the same ball park. An interesting result is noted with the CPU time for the run - it is very similar to the previous run that had only 3 elements. This is because with so many elements, the MPA converges very quickly because the maximum edge order is lower. This does not happen always. As mentioned previously, your best bet is still to use the AutoGEM defaults unless you have a really good reason not to (like no convergence on 9 passes).

Create and show the result windows for the Von Mises stress and the deformation animation by opening the rwd file you saved previously. These are shown in Figures 11 and 12 below. They are substantially the same as for the previous model. Increasing the number of elements with the AGEM settings has not really affected our results. Leave the results window and return to the model.

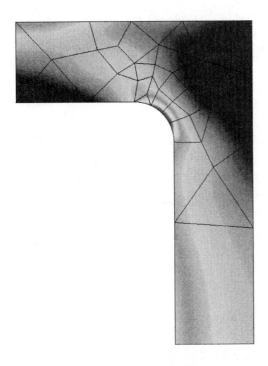

Figure 11 Von Mises stress with new AutoGEM settings

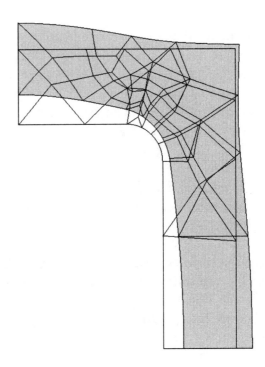

Figure 12 Deformation with new AutoGEM settings

Changing the Mesh with Detailed Fillet Modeling

As you probably know, fillets are used to reduce the stress concentration that occurs at sharp corners. Because stress gradients in the region of a fillet are usually high, it is necessary to exercise some care in setting up the model. MECHANICA offers a built-in option that will create a denser mesh in the vicinity of a fillet, without changing the global default AutoGEM settings on element size, edge turn, aspect ratio, and so on. This option is *Detailed Fillet Modeling*. Since we have a fillet in the current model, let's try this option out to see how it works.

In the pull-down menu, select

> *AutoGEM > Settings*

In the AutoGEM dialog window, check the boxes beside the following options:

♦   **Detailed Fillet Modeling** This is the main option we want to explore here.
♦   **Delete Mesh Points When Deleting Elements** While we play with the mesh, we don't want each mesh to leave remnants behind that interfere with the default action of AutoGEM.
♦   **Modify or Delete Existing Elements** We want AutoGEM to start from scratch each time we ask it to create a new mesh.

In the **Limits** tab, click the *Default* button to return all settings to their original values.

Open the **Analyses and Design Studies** window, and check the settings for the study **axitank**. Make sure the box **Create Elements During Run** is checked. Go ahead and **Start** the run. Delete the old files and accept error detection.

Open the **Study Status** window. The model contains 33 elements. The run converges on pass 5 with maximum edge order 5. The maximum X translation is 1.83E-04, and the maximum Y translation is 2.63E-04. Compare these to the previous values - pretty close! The maximum Von Mises stress is 4957 psi.

Open the **Results** window and use *File > Open* to load the **axitank.rwd** file we made previously. Show the Von Mises stress window (Figure 13) and notice the concentration of small elements created along the fillet.

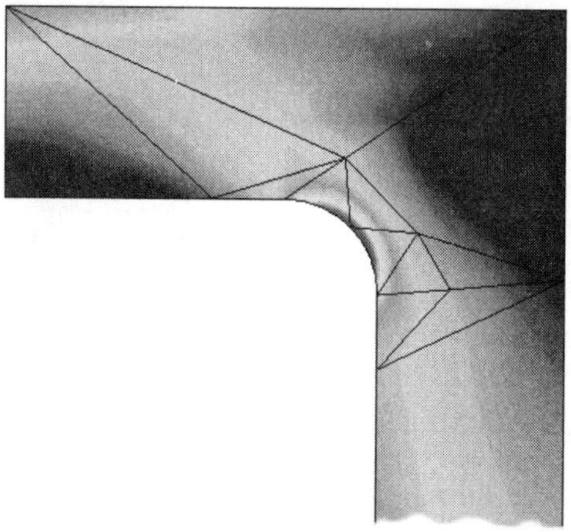

Figure 13 Von Mises fringe plot showing mesh created using detailed fillet modeling

Although using detailed fillet modeling did not have a significant affect on this model, you might investigate to see what happens if the fillet radius is much smaller (see Exercise #2).

Detailed fillet modeling produces a lot of elements in the immediate vicinity of any and all fillets in the model. It may be that only one or two of these are critical. So, how can we increase the local density of the mesh near critical fillets without changing global parameters or using detailed fillet modeling everywhere? There are a couple of ways to do this, including a new tool introduced in Wildfire 2.0.

Methods for Controlling the Mesh

The AutoGEM settings change grid creation parameters throughout the mesh (min and max edge angles, aspect ratio, edge turn, and so on). The detailed fillet modeling option affects fillets throughout the model. If we want to be a bit more selective about how the mesh gets modified, there are a couple of methods available. The first of these is a new function introduced in Wildfire 2.0.

Go to the pull-down menu *AutoGEM > Settings*, and make sure that detailed fillet modeling is turned off, and that mesh creation limits are reset to the default values. To make sure we are getting the default three elements, use the *Create p-mesh* button in the toolbar. We should get the mesh shown in Figure 9. Close the AutoGEM window and do not save the mesh.

In the right toolbar, select the *Create AutoGEM Control* button (also available in the pull-down menus with *AutoGEM > Control*). This opens the dialog window shown in Figure 14. The two types of control available with this control dialog are to specify the number of element nodes along specified edges (including variable spacing) and to tell AutoGEM to ignore edges smaller than a specified value. We will investigate the first of these, which is the default **Edge Distribution**.

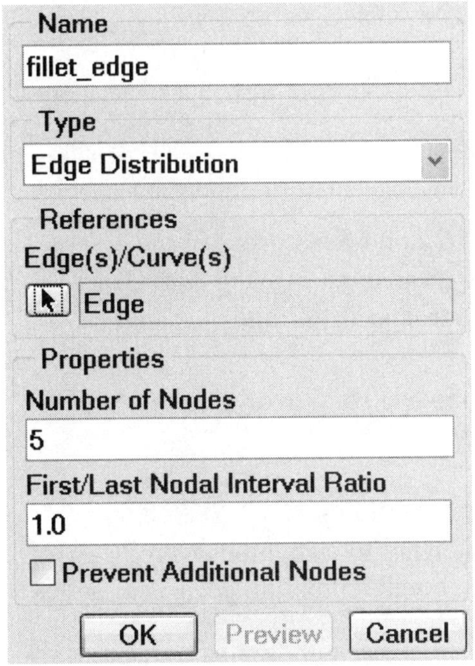

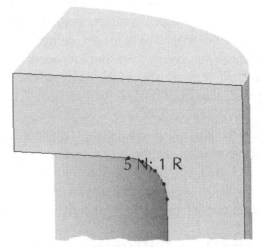

Name

fillet_edge

Type

Edge Distribution

References
Edge(s)/Curve(s)

Edge

Properties
Number of Nodes

5

First/Last Nodal Interval Ratio

1.0

☐ Prevent Additional Nodes

OK   Preview   Cancel

Figure 14 AutoGEM Control dialog window

Figure 15 AutoGEM control on fillet edge

Pick on the fillet edge and set the number of nodes to *5*. Enter the name of the control (that will appear in the model tree) as *fillet_edge*. Leave the nodal interval ratio setting at 1.0 (this allows you to set up a variable spacing between nodes; a value of 1.0 means equally spaced). With the dialog completed, select *OK*. The notation shown in Figure 15 appears on the model (N=5 nodes, R=1 interval ratio).

Go to the **Analyses & Design Studies** window, open the **Settings** display, and make sure the box **Create Elements During Run** is checked. The study should still be set to an MPA, with 5% convergence, maximum order 9. Run the study and open the **Study Status** listing.

AutoGEM has created 15 elements (a bit more than detailed fillet modeling). The run converged on pass 5 with order 5. The maximum X displacement is 0.0183 inch, the maximum Y displacement is -0.0263 inch, and the maximum Von Mises stress is 5011 psi.

Use your previously saved rwd file to display the Von Mises fringe plot, and the deformed shape.

You might like to create a result window showing the p-level fringe plot. This will show that most of the elements around the fillet were actually fairly low order, while the highest order elements were farther away, where the stress gradients were not as large but extended over longer element edges. Obtaining convergence using low order elements in the high gradient area of the model gives us some confidence about the results obtained.

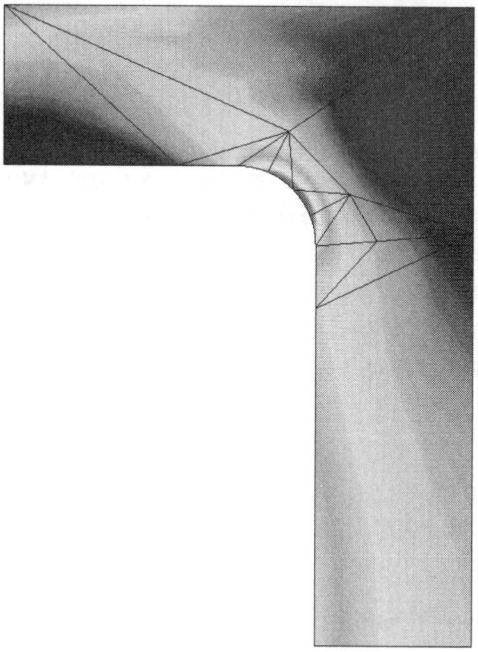

Now we will check out another way that can be used to control the mesh in a specific area of the model. It is a bit more flexible than node control

Figure 16 Von Mises stress fringe plot showing mesh created using AutoGEM mesh control on fillet edge

along the model edge, but adds additional complexity to the model. This involves creation of datum curves and nodes within the region to be meshed.

Use the **Sketch Tool** in the right toolbar to create the curve shown in Figure 17. Sketch this on the front surface of the model. Use the **Offset** command in Sketcher to create the arc. When you leave Sketcher, this will show up as a blue line. Now, create a couple of datum points along this curve as shown in Figure 17 (these are located at a ratio of 0.33 and 0.66 of the curve length, respectively).

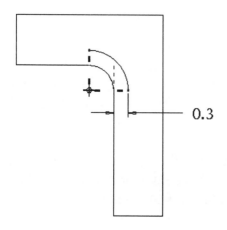

Figure 17 Datum curve to control mesh generation

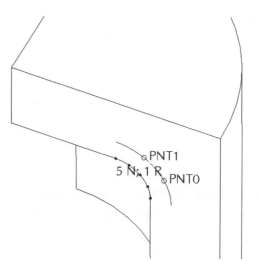

Figure 18 Datum points placed on curve to act as mesh "seeds"

Run the MPA analysis using this new control (keeping the previous edge control). The results are essentially identical to the previous results. The mesh (10 elements) is shown in Figure 19. Notice that nodes are placed at the ends of the datum curve, plus at the two datum points. Also, observe that the mesh contains both quad and triangular elements. Where are the high order elements here? Go to the AutoGEM settings, and direct AutoGEM to use triangular elements only. This results in the mesh shown in Figure 20. Results are again essentially the same as before. Where are the high order elements? What effect has the mesh had on CPU time?

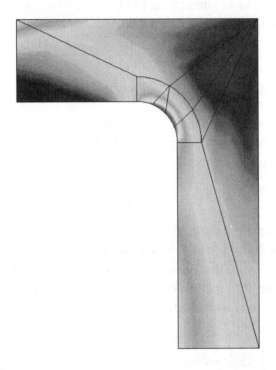 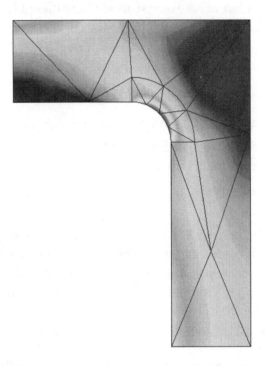

Figure 19 Mesh created using edge control plus datum curve and points (mixed mesh)

Figure 20 Mesh created using edge control plus datum curve and points (tri mesh only)

In summary, we have now seen several different methods for mesh control:

♦   AutoGEM settings - limits (edge angles, aspect ratio, etc.)
♦   Detailed fillet modeling
♦   AutoGEM Control for the number of nodes along an edge
♦   Datum curve and points for locating element nodes
♦   Selection of mesh geometry (triangles, or mixed quads and tri's)

For this simple model, these do not seem to have significantly affected the results - the results are certainly within our 5% convergence tolerance. The variation would be reduced if we used a tighter convergence criterion, but at the expense of increased CPU time. For this model, going an extra p-loop pass to obtain tighter convergence is not an issue. However, for much larger problems, the time required for the last pass can be a significant portion of the total. If that runs to many minutes (or even hours if you have minimal hardware), you will have to judge whether it is worth it!

These results should again give you confidence in using the mesh creation defaults. However, you will eventually run across a problem where you will want to have more active control over the mesh geometry. A typical example would be an area of the model where stress gradients are very high, and the use of large elements means that convergence cannot be obtained even with the maximum order set to 9. It is easy to get AutoGEM to create more (and smaller) elements in just those areas.

It is left as an exercise to see how these ideas might extend to controlling the mesh in 3D. For example, what role do points, curves, and surfaces have in shaping the mesh produced by AutoGEM.

Comparing to a Solid Model

Before we leave this tank model, let's compare results and performance with a solid model using the full 1/8 symmetric model created in Pro/E. This will also give us a chance to practice creating constraints for symmetry conditions. In the pull-down menu, select

Edit > Mechanica Model Type

Change to a **3D** model. When you leave this menu, you are informed that all modeling entities will be deleted. *Confirm.* Open the model tree to see the old loads and constraints are gone. We do not need the simulation features (the sketched curve or datum points), so you can delete them (right click in the model tree). We will have to recreate our constraints and loads. In dealing with the 3D solid, we must use constraints on three faces of the model that arise from symmetry about the horizontal plane and the two vertical datums.

Before proceeding, reset the AutoGEM settings to the default values.

Now we will set up the solid model. Select (or use the **New Displacement Constraint** button)

Insert > Displacement Constraint

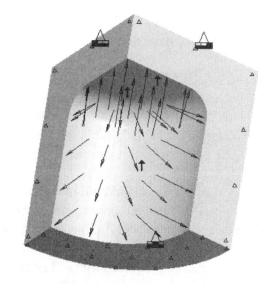

Name the first constraint *[XYface]* in **ConstraintSet1.** Select the Surfaces button and click on the front face of the model (in the XY plane). Middle click. For this surface, we must set the translation in the Z direction to **FIXED**, with the other degrees of freedom **FREE**. Recall that rotations have no affect on solids. Select *OK*. Continue on with the other surfaces (with the toolbar button or) with

Insert > Displacement Constraints

Name this constraint *[YZface]* (still in ConstraintSet1). Pick the left vertical surface. This surface must be **FIXED** against X translation with the other translations FREE.

Figure 21 Solid model with pressure load

Finally, create a third constraint on the lower surface of the solid called *[XZface]*. This must be fixed against translation in the Y direction.

Now we can apply the pressure load. Select (or use the **New Pressure Load** toolbar button):

> *Insert > Pressure Load*

Name the load *[pressure]* in **LoadSet1**. The default reference is now a surface. Pick on the three inside surfaces of the tank (including the fillet). Enter a magnitude of **1000**.

Now assign the Material **STEEL** to the part. The completed model is shown in Figure 21. Examine the constraint symbols very carefully.

Create a new static analysis called **axitank_solid**. Run the usual Quick Check and then MPA on this model. Use a maximum order 9 and 5% convergence. AutoGEM will create 34 solid elements. For the final MPA run, results are

> convergence on pass 6, max edge order 7
> max_disp_x   1.83E-04   (0.000183 in)
> max_disp_y   2.62E-04   (0.000262 in)
> max_stress_vm   4958 psi

Compare these results to the previous axisymmetric results. They are all within a few percent of each other. Notice that the total CPU time is several times (perhaps 10 or so) as long as that for the axisymmetric models.

Create result windows to show the convergence of the Von Mises stress and strain energy (Figure 22), the Von Mises stress fringe plot (Figure 23) and the deformation animation (Figure 24).

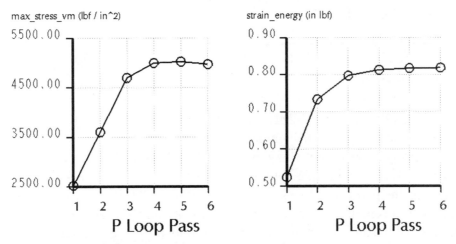

Figure 22 Convergence of Von Mises stress (left) and Total Strain Energy (right) for solid model

The convergence plots are about as perfect as we could wish. In the Von Mises fringe plot, look for axisymmetry in the fringes. Obviously we should see the same fringe pattern and values on both vertical faces. In the deformation animation, observe that the desired displacements are occurring (that is, deformation is consistent with the symmetry condition).

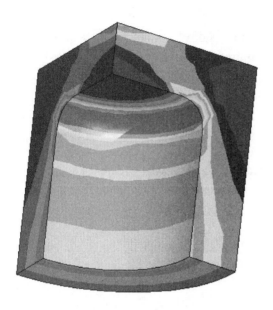

Figure 23 Von Mises stress in solid model

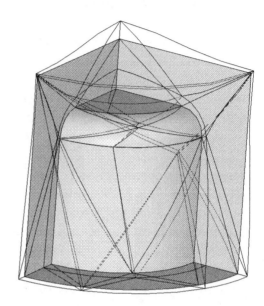

Figure 24 Deformation of solid model

The thing to note here is how much CPU time was saved by using the axisymmetric model, without significantly affecting the accuracy of the results.

We will continue on, now, with another idealization that involves axisymmetric models. Leave MECHANICA and return to Pro/E. Remove the axitank model from the session. This is also an appropriate place to take a break!

Axisymmetric Shells

This example will illustrate a number of MECHANICA functions we haven't seen before: using 2D Shell elements and applying a centrifugal load. The problem involves the analysis of the hollow axisymmetric object shown cut-away in Figure 25. We'll call it the centrifuge model. The wall thickness is very small compared to the overall dimensions, so we will use shell elements. Loading on the part is due to a high speed rotation (3000 RPM) about the symmetry axis. This produces an acceleration at the outer rim of about 2000 g's (1g = 9.81 m/s^2). We are interested in finding out the stresses in the material, and how these are affected by the location of the vertical interior cross brace. As an alternative to the interior brace, we will also see the effect of pressurizing the interior of the model. To do this, we must use a round-about method of applying a pressure load to axisymmetric models, since this is not available in integrated mode.

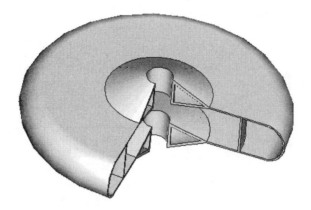

Figure 25 The centrifuge part - 400mm diameter,
3000 revolutions per minute

Creating the Model

The interesting thing about this idealization is that no solid part is required! As for the previous axisymmetric solid, we only need a 2D representation of the revolved cross section. Furthermore, all we need to create (in Pro/E) is a set of datum curves that represent the walls in a cross-sectional view of the model[4]. These curves are used by MECHANICA to create *2D Shell* elements, which represent what would happen if the curves were revolved through 360° to form surfaces. The only thing we must be careful about is to create the defining curves in the XY plane, where the Y axis is the axis of revolution of the model. Also, due to symmetry about the horizontal plane, we only need to create curves for the upper half of the centrifuge.

Start a new Pro/E part called **[centrifug]**. Use the **mmNs_part_solid** part template. The geometry of the model is shown in Figure 26, minus the vertical brace which we'll add later. Each thick line is a datum curve. All curves are on the **FRONT** datum. This geometry can be created in a couple of ways:

♦   Create a pattern (using a pattern table) of datum points at the nodes and then join pairs of points using datum curves. This is a bit laborious!
♦   Create the curves all at once as a single sketched curve - probably the easiest way, and the one used here. You don't need a datum axis for the axis of symmetry - the Y-axis is automatically assumed.

Go ahead and create the curve shown below using the **Sketch Tool**. Create the curve on the FRONT datum plane, and note the X and Y directions of the default coordinate system.

---

[4] Axisymmetric shell models do not have to be created from datum curves. If we had a solid model, we could use physical edges (formed by cutting the solid on the XY plane) as the geometry references to define the model.

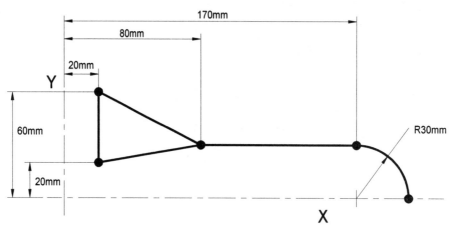

Figure 26 Dimensions for the axisymmetric half-model, without the vertical internal brace

Setting the Model Type

When you have the sketched curve finished, go into MECHANICA with

Applications > Mechanica

Make sure your model units are correct (mm - N-s). In the Model Type window, select *Advanced*. Then pick the option *2D Axisymmetric*. Pick the selection button (under **Geometry**) and then pick on the sketched curve - it will highlight in red. Middle click. Then pick the *Coordinate System* button and pick on the default model coordinate system. Leave the Model Type menu with *OK*. The curves now are highlighted in purple.

Setting Constraints

We need to constrain the part against moving in the Y direction only. We will constrain the point on the horizontal centerline (due to symmetry, the vertical displacement of this point must be zero) and also the vertical element at the hub (parallel to the axis). In the right toolbar, select

Displacement Constraint

Call the constraint *[endpoint]*. We will use **ConstraintSet1**. Select the **Points** option under References, and pick on the endpoint where the arc segment meets the X-axis, then middle click. Set the X translation **FREE** and leave the Y translation and Z rotation **FIXED**. (Why fix the rotation here?) See Figure 27. Accept the dialog with *OK*.

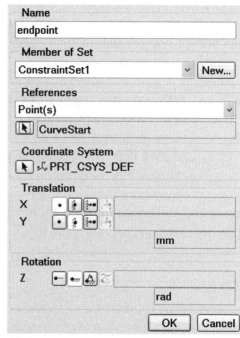

Figure 27 Constraining a point

Now constrain the left vertical edge (the hub):

> *Displacement Constraint*

Name the constraint *[hub]* (also in **ConstraintSet1**). In the References list, select the Edges/Curves option and select the vertical edge at the left end. You will have to use a right click to preselect just the single vertical edge, then left click to accept it. Middle click. This will be completely constrained (all degrees of freedom **FIXED**). Accept the dialog with *OK*.

Setting a Centrifugal Load

To apply a centrifugal load is quite easy. The axis of rotation for 2D axisymmetric models is always the Y axis. All we need to specify is the speed of rotation in radians per second. Use the toolbar button for **New Centrifugal Load** or select

> *Insert > Centrifugal Load*

Enter a load name *[cent314]* in LoadSet1. Enter a magnitude for the angular velocity of *314* rad/sec, which corresponds to 50 rev/sec (3000 RPM). The other dialog areas are grayed out at this time. For an axisymmetric model, the axis of rotation is always the Y axis (note the direction vector components < 0, 1, 0 >). For general 3D models we can specify the axis of rotation using a vector direction defined by three components points. Accept the dialog. Note the green centrifugal load symbol on the Y axis (this is easier to see if you turn off coordinate system display).

Setting Shell Properties

To complete our model, we need to specify the shell wall thickness and the material. We will use two different shell thicknesses in this model. Thickness and material are both shell properties, and are accessed using (or use the **New Shell** toolbar button)

> *Insert > Shell*

Enter a name for this definition *[thick5]*. Select the **Edges** button. Because all the edges belong to the same datum feature, we can't just pick on the desired segment - the entire feature would be selected. We must be a bit more choosy here.

Instead, put the cursor on the horizontal segment and right click. (This is like the old **Query Select** mode in Pro/E.) Just that segment will highlight in blue. Now

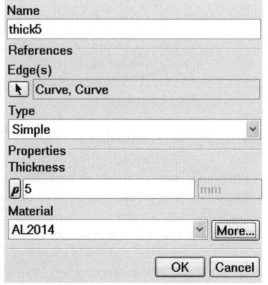

Figure 28 Shell property definition (thickness and material)

left click and the segment will highlight in red. Right click again on the arc segment. It highlights in blue. This time, you must hold down CTRL when you left click on the arc. It will also highlight in red, and our selection set now contains both the curve segments. Middle click.

Enter a thickness of **5** (recall we are using mm). Select the ***More*** button and move the material **AL2014** over to the model list on the right. Use ***OK*** to accept the selection (what happens if you use ***Close*** instead?). The complete shell property definition window is shown in Figure 28. Accept the definition with a middle click. The shell elements are shown in green.

Repeat the above procedure to create another property definition ***[thick8]***, and assign this property to the three curves forming the triangle at the hub. The thickness is 8 mm and the material is AL2014 as before. The model is now completed (Figure 29).

Open the model tree to see the shell definitions. In the model tree, right click on Loadset1, and select ***Rename*** in the pop-up menu. Change the name to ***[centrifugal]***. We will be adding a second load set for the pressure load a bit later.

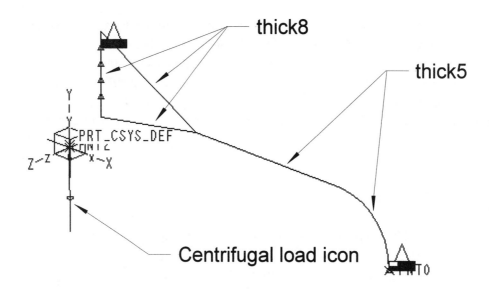

Figure 29 The completed axisymmetric model (isometric view)

Running the Analysis

As usual, the first time we run a model, we will perform a Quick Check to see if there are any serious modeling errors. Start in pull-down menus:

Analysis > Mechanica Analyses/Studies > File > New Static

Enter an analysis name ***[centrifug]***. Enter a description and make sure the defaults **ConstraintSet1** and **centrifugal** are highlighted. Finally, select **Quick Check** and leave the dialog.

Check the *Run Settings* and select the desired directories for output and temporary files. Then select *Start*. Accept error detection. Check the **Study Status** window for any reported errors or warnings. There are 5 *2D Shell* elements. You might note the maximum Von Mises stress is 9.4 MPa; maximum X and Y deflections are 0.014 mm and -0.019 mm, respectively.

Since there are (or should be!) no errors, we can change the analysis. In the **Analyses and Design Studies** window, highlight the study and select (in the RMB pop-up):

> *Edit*

Change to a multi-pass adaptive analysis with **5%** convergence on **Local Displacement, Local Strain Energy & Global RMS Stress**, a maximum polynomial order of **9** and accept the dialog.

> *Start Run*

Delete the existing output files for the model. It is probably a good idea to always use error detection. Open the **Study Status** window. The analysis converges on pass 3 with a maximum edge order 5. The maximum Von Mises stress has increased to 11.2 MPa; maximum X and Y deflections have also increased to 0.0185 mm and -0.0216 mm, respectively.

View the Results

Create the usual result windows for the Von Mises stress and a deformation animation. Call the first window *[vm]*. Get the output directory **centrifug** from the location you specified under **Run Settings**, enter a window title, set the **Quantity (Stress, Von Mises)**[5]. Select *OK and Show*. Using the top toolbar button, *Copy* the window definition to a second window called **[deform]** and modify the definition to set up a deformation animation.

Make the von Mises stress window active (yellow border) and select

> *Info > Model Max*

A label is placed on the model to show the location and value of the maximum stress. See Figure 30. If you use *Info > Dynamic Query*, you can determine stresses at other points in the model, but this is a bit tricky since you must be very accurate in picking the plotting grid points.

---

[5] You will have to use the Fringe display type. In independent mode, there is another type of display called Query that lets you display stress at each point in the plotting grid.

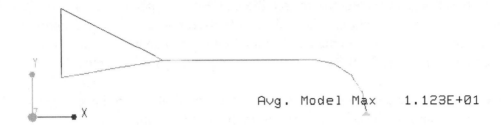

Von Mises Stress

Figure 30 Von Mises stress in axisymmetric shell model with centrifugal load

Figure 31 shows the (exaggerated) deformed shape. What is the displacement scale? Recall that the maximum Y displacement is only 0.02mm.

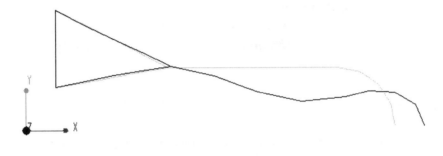

Deformation

Figure 31 Deformation of the axisymmetric model with centrifugal load

From the deformation we see that due to the high centrifugal load, the wall of the shell is drawn inwards. We will add a vertical brace inside the shell to stiffen it in the Y direction. Before you leave the **Results** window, save the rwd file for this study (call it **centrifug.rwd**).

Modifying the Model

For all our models up to this point, we have created the geometry in Pro/E and brought it into MECHANICA. We are going to add to this geometry in MECHANICA to create what might be called a hybrid model. It is important to realize that the simulation feature we add here is not known to Pro/E and will disappear when we leave MECHANICA[6]. In fact, it would be possible to create this entire model within MECHANICA using simulation features alone. This only applies here because there are no solid features in this geometry. Those cannot be created in Pro/M.

---

[6] Unless it is "promoted". See the RMB pop-up for the feature after we are finished.

In the right toolbar[7], select

Sketch Tool

Pick on **FRONT** for the sketching plane, and **TOP** as the top reference. In Sketcher, pick the horizontal datum curve as a reference. Create a vertical line at X = 130 between the X-axis and the horizontal element. See Figure 32. Accept the sketch and observe the color of the line we just created. It is different because it is currently not in the model. Go to

Edit > Mechanica Model Type

Select the button under Geometry references, and CTRL-click on the new curve. Middle click to accept. It is now included in the model.

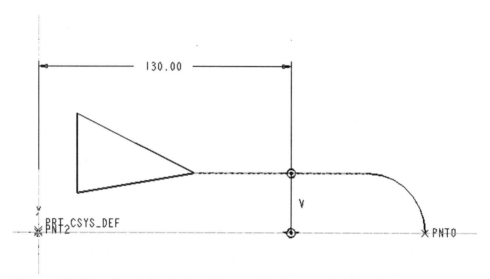

Figure 32 Sketch of simulation feature for the vertical brace

We need to apply properties to this new geometry item (we'll just add this curve to the **thick8** group). Select one of the three current edges in the **thick8** group, then in the RMB pop-up, select **Edit Definition**. Select the **Edges** button, and then CTRL-pick on the new datum curve. Middle click and then select **OK**. The shell set **thick8** now contains four curves. If you Repaint or reorient your screen, note that the new curve for the vertical brace is shown in a different color than the curves imported from Pro/E.

We need to add a symmetry constraint on the lower point on the new brace. This is easiest to do by adding the endpoint of the new curve to the existing constrained point definition. Do that by selecting the constraint, then in the RMB pop-up, select Edit Definition. Add the new curve end point. The new model should look like Figure 33. Open up the model tree and expand all the branches to see the data structure of the model.

---

[7] Alternate sequence: highlight FRONT, select the **Sketch Tool**, middle click

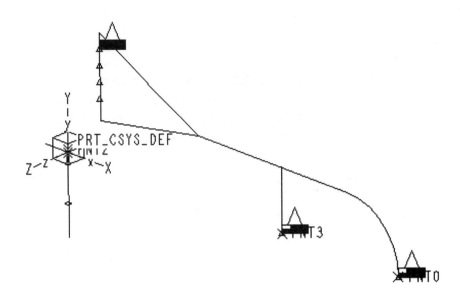

Figure 33 Axisymmetric model with additional brace (isometric view)

Running the Modified Model

Open the ***Design Study*** dialog. We shouldn't have to make any changes to the analysis type, so we can go directly to

Run Settings

and check your settings. Make sure "Create elements during run" is selected (since we have added a new shell). ***Start*** the analysis and open the **Study Status** window. Note that there are now 7 elements - the horizontal curve has been split at the vertex/junction with the vertical brace. The run converges in 3 passes with a maximum edge order of 5. The maximum Von Mises stress is reduced from the previous value of 11.2 down to 8.99 MPa. The deflections are now 0.0146 mm and -0.0097 mm in the X and Y directions, respectively. The Y displacement has been cut in half from the previous model.

Create the same result windows as before (easy to do if you have the rwd file stored!). Open the window showing the Von Mises stress and use ***Info > Model Max*** to have a look at where this occurs. Results should be as shown in Figure 34.

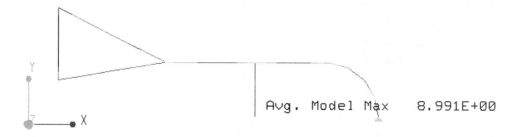

Avg. Model Max   8.991E+00

Von Mises Stress

Figure 34 Von Mises stress in modified model

The deformation of the modified model is shown in Figure 35. What is the deformation scale? The vertical brace has partially prevented the collapse of the side wall, as intended, and also served to reduce the maximum stress in the model, which still occurs at the outer circumference.

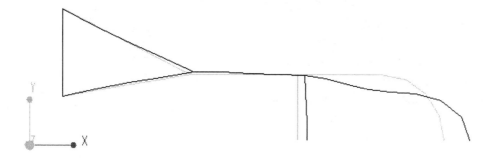

Deformation

Figure 35 Deformation of the modified model

It would be interesting to find out how the location of the vertical brace might affect the stress and deformation in this model. From Figure 35, it looks like moving it farther out would reduce the vertical collapse even more (except remember that the "centrifugal" load increases in proportion to r^2, so this might increase the stresses in the hub). This problem is left as an exercise.

Pressure Loads on Axisymmetric Shells

Another way that might reduce the sidewall collapse is to pressurize the inside of the centrifuge.

First, in the model tree, select the simulation feature (the sketched curve) and delete it.

Applying pressure loads on axisymmetric shell models in new in Wildfire 2.0. Pick the *New Pressure Load* in the right toolbar (or in the pull-down menu). Create a new load set called *[Pressure]*, to keep it separate from the centrifugal load. Name the load *[pres100]*. The References are automatically going to be edges. We will select two edges, but they must be in separate loads for reasons we will see shortly. Select the horizontal edge of the model. Middle click, then enter a value of **0.1** (units N/mm² = MPa; 0.1 MPa is about 1 atmosphere). If you select *Preview*, you will see the load arrows pointing downward - this is a consequence of which way MECHANICA is considering the positive normal direction to the edge. We want the pressure to act upwards on the shell, so change the load value to **-0.1**. Select *OK*. The load arrows will show in yellow.

Select the *Pressure Load* toolbar button again. Keep the load in the set **pressure**. Name this load something like *[end_arc]*. Select the arc on the end of the model. Enter a value of **0.1**. Select *Preview*, and enter a negative value if necessary to have the arrows pointing outward. Select *OK*.

To make sure our pressure load is going the right way[8], in the top pull-down menu select

Info > Review Total Load

Pick the **Select Loads** button and select the pressure load on the arc. Then select *Compute Load Resultant*. The resultant of this pressure load is indicated to be 3487.17 in the positive Y direction. Why is there no resultant force in the X direction? Can you confirm that this is the correct value? The important thing here is that the resultant is upwards, despite what the display arrow may be telling you!

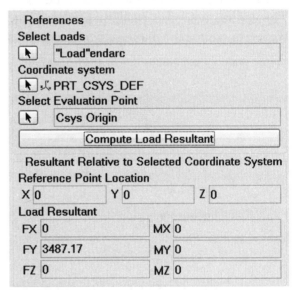

Figure 36 Calculating the Load Resultant for the pressure load on the arc

Go to the Design Study window. Copy the study centrifug to create a new study centrifug_total. Edit the new study to make sure both load sets (centrifugal and pressure) are selected. Go straight to an MPA convergence analysis. Start the run, remove existing files, and open the **Study Status** window. The run will converge on pass 3 with a maximum edge order of 5. For the centrifugal load, the maximum von Mises stress is 11.2 MPA (as before). For the pressure load set, it is 26.9 MPa. Note that for the pressure load, the maximum displacements in the X and Y directions are in the opposite directions from the centrifugal load.

Create separate result windows for the deformation due to only the centrifugal and pressure loads. When you create these windows, set the scale factors for each load set to either 1 or 0 in order to isolate the two loads. Set the deformation scales in each of these windows to 200. See Figure 37. Observe the direction of the pressure arrows. Clearly, the pressure is pushing the sidewall outward.

---

[8] In the first production release F000 of Wildfire 2.0, there is/was a discrepancy between the direction shown with the Preview function (correct), and the display of the load arrows in the normal model display (backward).

Figure 37 Deformation due to 0.1 MPa pressure load (left) and 3000 RPM centrifugal load (right). Scale factor 200.

Create a result window showing the two loads acting simultaneously. Set the scale factor on the pressure load to 0.3. This puts an internal pressure of 30 kPa (about 4.5 psi) inside the centrifuge. Copy the window definition for the combined loading and edit it to show the Von Mises stress. The maximum von Mises stress is now 8.6 MPa (this is less than we had with the vertical brace).

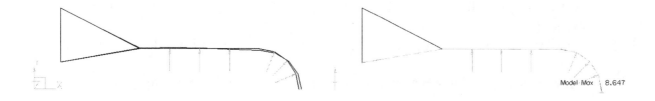

Figure 38 Deformation (left) and Von Mises stress (right) for combined loading of 3000 RPM and 30 kPa internal pressure. Maximum stress 8.6 MPa.

By keeping separate load sets for the centrifugal load and the pressure, we could conceivably find what pressure would be optimal (produce minimum stress levels) for several possible values of the centrifugal load. Here is a question for you to investigate: If we double the scale factor on the centrifugal load, does this mean doubling the rotation speed of the centrifuge, or some other factor?

Summary

Axisymmetric models are quite common, and you should be familiar with both solid and shell elements. Fortunately, these are easy to set up and the computational load is very light, so the models will execute quickly. Axisymmetric models can also combine solid and shell elements (consider a flywheel with a thick inner hub and outer rim connected by a thin disk). We will see one of these combined models in the next chapter. You should try to do some of the exercises below to get more practice at creating and interpreting these models.

Questions for Review

1.   On what plane do you have to create an axisymmetric model? Where is the axis of symmetry? How is this indicated in the model?

2.   What types of elements are available for axisymmetric models? Make some quick sketches or describe typical shapes appropriate for each element type.

3.   What are the main load types available for axisymmetric models? How is the MECHANICA definition different from standard FEM package definitions?

4.   What is the minimum constraint required in an axisymmetric model?

5.   What are three ways you can modify the default mesh created by AutoGEM?

6.   Is it possible to have an axisymmetric model which, under load, will deform such that it crosses the Y-axis?

7.   At what point in the model creation do you identify it as a 2D axisymmetric model?

8.   At various times, sketched curves may be displayed in red, blue, purple, or green. What is the significance of these colors?

9.   How can you apply pressure to an axisymmetric shell?

10.   What will generally happen to the convergence when you produce a finer mesh in MECHANICA? Why?

11.   What will happen to the mesh created by AutoGEM if you sprinkle datum points on the model?

12.   How do the usual AutoGEM limits on edge turn work on datum curves placed on the model?

13.   Is it possible to create 2D Solid elements that overlap? What about 2D Shell elements? First, do you think this is a reasonable thing to be able to do, and second, find out how MECHANICA will respond if you do this.

14.   Find out if it is possible to create shell elements of varying thickness along an individual curve segment. If not, how would you approach modeling this?

15.   What is the *lowest* edge order used in a multi-pass adaptive analysis of a 2D Shell model?

16.   When specifying a centrifugal load, what units are used for the rotational speed?

17.   What happens if you apply a gravity load in the X direction in an axisymmetric model? (This is a trick question!)

18.   How are the material and thickness specified for a 2D Shell?

19.   How is the material specified for a 2D Solid?

20.   Why is no radial constraint required in an axisymmetric model?

Exercises

1. For the axisymmetric solid model of the steel tank, we fixed the Z rotation along the symmetry edge. Why? Remove this constraint and comment on the results.

2. For the axisymmetric tank model, what is the effect of the fillet radius on the maximum von Mises stress? Investigate this using a sensitivity study with the radius varying from 0.1 to 1.0 inch. Try this with and without detailed fillet modeling and comment on the results. In particular, what is the effect of detailed fillet modeling for large radius fillets?

3. Consider a 3" diameter circular steel rod with a circumferential semi-circular groove under an axial tension load.
    (a) Assuming that the load is 100 lb, find the maximum axial stress in the rod if R=0.25".
    (b) Do a sensitivity study for values of R ranging from 0.10" to 0.50" in order to find the variation in the maximum axial stress as a function of R. Plot these results in terms of a stress concentration factor, with the axial stress normalized with a nominal stress based on the minimum rod diameter. Compare this factor with published values (you'll find these in mechanical design textbooks). Justify your choice for the dimension L. HINT: You may have to play around with the mesh to get reliable results. Comment on the use of FEA for obtaining stress concentration factors.

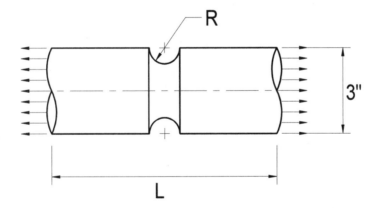

4. In our centrifuge problem, we want to find out the effect of the location of the interior brace. This calls for a sensitivity study. Set this up with the location dimension to the curve in MECHANICA as the design parameter. Vary this distance from 100 to 160 mm. Plot the variation of the maximum Von Mises stress and maximum displacement magnitude with this parameter.

5. As suggested in the text, find the relation between the load scale factor for the centrifugal load, and the specified rotation speed in the load definition. That is, if the scale factor is doubled, what is the effective change in speed?

This page left blank.

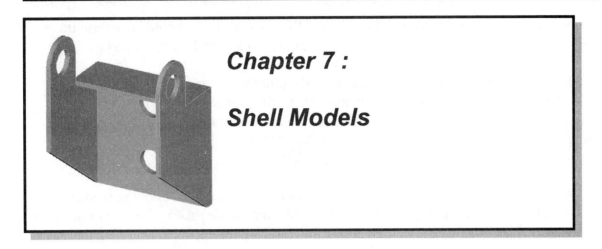

Chapter 7 :

Shell Models

Synopsis

Shell models for general 3D geometry; automatic detection of shell surfaces; manual shell setup; reentrant corners in shell models; mixed solid/shell models; constraints in cylindrical systems; bearing loads

Overview of this Lesson

After the default solid elements, the next most common element for 3D modeling is the shell. In this lesson we will investigate idealizations using shell elements to represent thin-walled solids defined by pairs of parallel surfaces. In the previous lesson, we saw that an axisymmetric shell model could be constructed using datum curves. The shell properties (thickness and material) were specified using a dialog window and then assigned to the geometric curves. In particular, the shell thickness was supplied as input data rather than coming from the Pro/E model itself. The same situation occurred with the plane stress model earlier.

For general 3D shell models, things are a bit different. In this case, MECHANICA reads the shell thickness directly from the Pro/E solid model. Shell models (or the portion of the model to be represented using shells) are determined by pairs of parallel planar or non-planar surfaces (that is, shells can be either flat or curved as long as the defining surfaces are parallel or concentric). The two defining surfaces in a pair are *compressed* to a mid-surface location where the shell elements will be created. In integrated mode, only shells of constant thickness can be treated. Independent mode allows the shell thickness to vary (ie non-parallel defining surfaces). The same model can have shells of different thickness formed by different pairs of surfaces, however. And, of course, the shell thickness(es) can be used as a design parameter(s) for sensitivity studies and optimization.

The purpose of shell models is to produce more efficient models. If portions of a solid model are composed of thin-walled features, treating them as solids is very inefficient (and sometimes prohibitive). The number of solid elements required to represent these features can be enormous. The general guideline for using a shell is that the thickness dimension should be less than about $1/10^{th}$ of the length of the shortest edge of the shell surface. It is possible to put shell elements on tightly curved corners (like fillets), but the radius of the fillet or round should be several times the shell thickness. We will investigate this in one of the models studied in the lesson.

We will look at three simple examples to illustrate the procedures for analysis. In the first example, the surface pairs forming the shell are determined automatically. In the second example, we will identify the pairs manually. This model will also illustrate a problem with obtaining convergence (unrelated to how we made the shells). In the final example, we will create a model containing both solid elements and shells.

An important part of this lesson is to see one of the important consequences of the p-code basis of MECHANICA. This involves the presence of a mathematical singularity that occurs at sharp interior corners. This is called a re-entrant corner and results in problems in assessing the convergence of the solution. The same effect occurs in solid models so, even if you don't plan on using shell elements, you should go through this anyway!

Automatic Shell Creation (Model #1)

Creating the Geometry

We will analyze the small pressurized tank shown in Figure 1 (approximately in default orientation). The tank has a hemispherical shape on the bottom. Create this new part called *[shelltank]* making sure that you set up the units as mm-N-s. The dimensions (in mm) of the major features of the tank are shown in Figure 2. Note that the round dimensions are given in Figure 1. Use the *Shell* feature in Pro/E to create a wall of uniform thickness (1.0mm) throughout. To take advantage of symmetry, create a final vertical cut through the model to remove the front half (as in Figure 2). Turn off all datums (planes, axes, etc.).

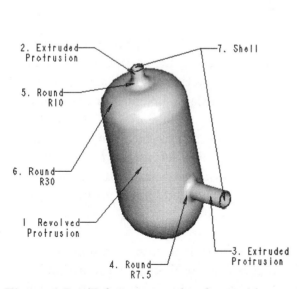

Figure 1 Pro/E features used to form tank

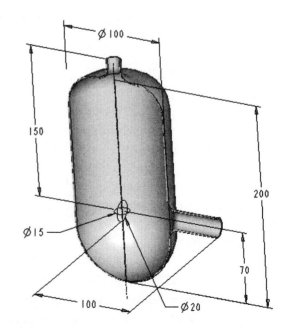

Figure 2 Tank dimensions (mm)

Defining the Shells

When the geometry is complete, select

Applications > Mechanica

Recall that the default model type is 3D, so we don't need to change that. Our first job is to define the shells in the model. In the pull down menu select (or use the *New Shell Pair* toolbar button)

Insert > Midsurface

then select *Auto Detect*. If shaded display is turned on, outer and inner surfaces will highlight in red and yellow; if shading is turned off, you will see yellow and red on the edges of the paired surfaces. See Figure 3. Then click on

Compress > Shells Only > ShowCompress

This will show the midsurface (yellow highlight) between the two model surfaces that will be used to create the shell elements (Figure 4). Note that *Show Both* displays the shell edges in yellow and the original surface edges in green. The shells are created at the midsurfaces of the pairs.

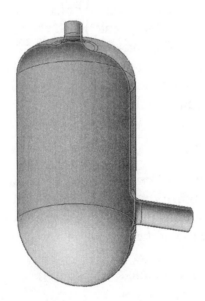

Figure 3 Automatically detected paired surfaces

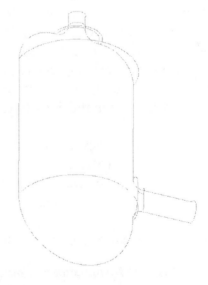

Figure 4 Compressed surfaces only

In the MIDSURFACES menu, select *Done/Return*. Open the model tree to see the shell entries that are listed under **Idealizations** - there should be 9 shell pairs. Selecting these in the model tree will cause them to highlight (in red and yellow) on the model.

Go to the pulldown menu and select *AutoGEM*. Notice that the option **Solid/Midsurface** is now checked. The next time you create a mesh for this model, AutoGEM will automatically use the paired surfaces to create shell elements on the midsurface.

While you are in the AutoGEM menu, make sure your AutoGEM limits are set to the default.

Assigning the Material

We now need to assign the usual materials, constraints, and loads to the model. Start with the material. In the right toolbar, select:

> ### *Define Materials*

Bring the material **SS** (a stainless steel) from the library into the model. Then select

> ### *Assign > Part*

Thus, material assignment is done in exactly the same way as for a solid model.

Assigning the Constraints

In 3D solid models, we normally apply constraints to surfaces. For the symmetry constraint in this model, that would be the thin surfaces created by the symmetry cut. However, those surfaces will disappear when the shell surface is created by compressing the surface pairs. So instead we will apply constraints to shell edges. We can pick either the inner or outer surface edge - both will collapse to the midsurface. Also, remember that, unlike solid elements, shell elements have rotational degrees of freedom. So, remember that for shell models:
1. we must apply constraints to edges or curves, and
2. we must keep rotation of those edges in mind.

Wildfire 2.0 contains a new constraint command (*Symmetry Constraint*) that is a shortcut for creating constraints involving either mirror or cyclic symmetry. As usual, we will forego this shortcut for a while to see what is actually involved in setting up the mirror symmetry constraint for the tank. You should come back later to explore the use of the this command.

In the pull-down menu, select (or use the **New Displacement Constraint** button),

> ### *Insert > Displacement Constraint*

Call the constraint *[symedges]* (member of **ConstraintSet1**). Set the **References** type to *Edges/Curves*, and go around the outer edge of the solid model on the symmetry (XY) plane and pick all the edges of the tank using CTRL to add edges to the selection set. You may have to zoom in to do this. Each edge will highlight in red when selected. When all edges are selected (there are 17 edges forming two tangent chains - note the counter beside the selection filter at the bottom), middle click. Symmetry requires that the Z translation (normal to the symmetry plane) be **FIXED**. The X and Y translations are both **FREE**. Also because of symmetry, we need to set the X and Y rotations as **FIXED**, and **FREE** the rotation around Z. Think carefully about these constraints (especially the rotations) and how they arise from symmetry. Accept the dialog.

We will also constrain the edges of the side and top inlet/outlet pipes. Select (or use the button)

Insert > Displacement Constraint

again. Name the constraint *[sidepipe]* (still in ConstraintSet1). Set the **References** type to *Edges/Curves* and pick the outer edge of the pipe coming out the side of the tank. Set all degrees of freedom to **FIXED** for this edge (no translation, no rotation). Repeat for the pipe leaving the top of the tank (call it **[toppipe]**). The constraints should appear as shown in Figure 5 (these are a bit of a jumble at the top of the model). Use the *Simulation Display* options to turn off the load and constraint labels.

Notice that the model is now overconstrained. That is, we have defined more constraints than are required to prevent rigid body motion. Can you list the redundant constraints? These are not "wrong" as long as they are consistent with our intentions for allowed deformation of the model. In this case, we are assuming that the two pipe connections to the environment are immovable and inflexible relative to the tank.

Assigning a Pressure Load

Now, in pull down menu, select (or use the **Pressure Load** button)

Insert > Pressure Load

Name the load *[presload]*, in **LoadSet1**. Select the button under Surfaces and CTRL-click on all the interior surfaces. Each surface (there are 8) will highlight in red as it is selected. Don't forget the interior surfaces of the rounds. You may have to spin the model to ensure that all surfaces are picked (it also helps if you are in shaded mode). Then, middle click. Enter a load magnitude of **0.1** (recall that our units for pressure are MPa; our applied pressure is equal to 100 kPa, about atmospheric pressure). Accept the dialog. The model should now appear as shown in Figure 5.

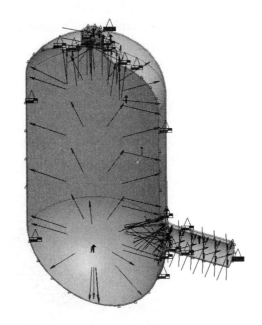

Figure 5 Completed model

Defining and Running the Analysis

We can now define the analysis:

Analysis > Mechanica Analyses/Studies > File > New Static

Enter a name *[shelltank]* and a description. Make sure constraint and load sets are selected. Select a **QuickCheck** convergence. Review the ***Run Settings***, and ***Start*** the analysis. Always accept error detection for the first run of a new model. Open the ***Study Status*** window. About 50 shell elements are created[1]. The maximum Von Mises stress is around 18 MPa and the maximum displacement is about 0.006 mm. Assuming no errors, change the analysis to a **Multi-Pass Adaptive** convergence (5% on **Local Displacement, Local Strain Energy & Global RMS Stress**, max order 9) and rerun the analysis. You can use the elements from the existing study. Open the ***Study Status*** window. The run should converge on pass 7 with a maximum Von Mises stress of about 17.4 MPa and a maximum displacement of 0.0091 mm. This is one of the few times you may see the stress decrease from the QuickCheck to the MPA analysis. Just above the reported measures, examine the values of the components of the resultant load on the model. Can you explain/verify these values?

Viewing the Results

Create some result windows for Von Mises stress and deformation animation. Show the element edges. Two views of the Von Mises stress are shown in Figure 6 below. Observe the location of the stress hot spots in relation to the deformation. This is easy to do if you show the stresses on the deformed shape. Note the scale on the deformation. Is the deformation consistent with your expectations? Locate the position for the maximum stress. Observe the variation in size and shape of the shell elements over the mesh. Create a p-level fringe plot and find out which elements have the highest edge order, and whether this automatically corresponds to the maximum stress level.

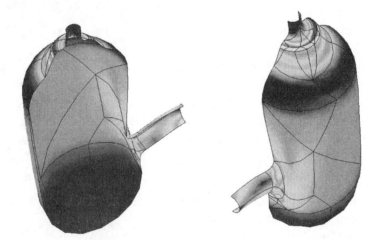

Figure 6 Von Mises stress fringe plot

The convergence behavior for this model is shown in Figure 7. This is a pretty well-behaved model.

---

[1] Different builds of Wildfire may produce slightly different meshes, and corresponding small changes in the results.

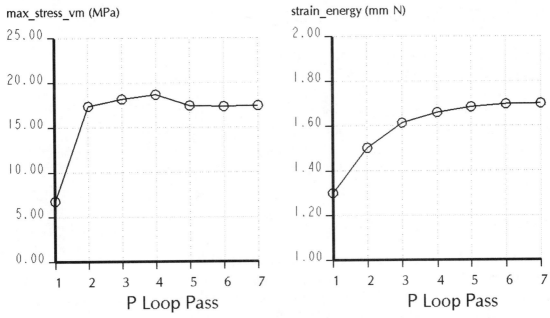

Figure 7 Convergence of Von Mises stress (left) and total strain energy (right)

Exploring the Model

You should spend some time exploring this model. Here are a few things to try:

1. Delete the existing constraint on the cut edges of the shell. Check out the new ***Symmetry Constraint*** command. This is a button available in the right toolbar (or use ***Insert > Symmetry Constraint***). The operation of the command is pretty self-explanatory. There are two types: mirror (default) and cyclic. We will look at cyclic symmetry a bit later. Select the same edges as before and re-run the analysis. Do the results using this constraint agree? What is the display icon that represents a ***Symmetry Constraint***?

2. In the ***AutoGEM Settings*** menu, there is an option for **Detailed Fillet Modeling**. Find out what this does and what effect it has on the results.

3. Back in Pro/E, change the radii of the rounds on the two pipes. How small can these be without significantly affecting the results? What happens if you suppress the rounds altogether?

4. Modify the constraints on the side and top pipes. Remove all rotational constraints. For the side pipe, specify only translation in the X direction as fixed. For the top pipe, specify only translation in the Y direction as fixed. Coupled with the symmetry constraint (translation Z fixed), these are sufficient to remove all rigid body degrees of freedom. What effect does this have on the results?

When you have finished your explorations, return to Pro/E and remove the tank model from the session. We will move on to our next model.

Manual Shell Creation (Model #2)

Once again, the model (Figure 8) is created in Pro/E and we will use an idealization of the solid to create shell elements. In the previous model, we let MECHANICA automatically detect suitable surface pairs to form shell elements. This does not always work. In this model, we will manually select the surface pairs that will be compressed to form the shell surfaces.

Creating the Model

Create the model *[bracket]* according to the dimensions shown in Figure 9. Make sure your units are set to mm-N-s. The view of the part in Figure 8 is approximately in the default orientation. Note that there will be two different shell thicknesses (5mm and 10mm).

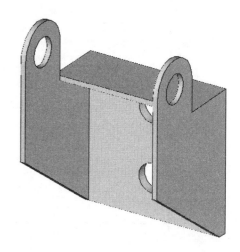

Figure 8 The mounting bracket model

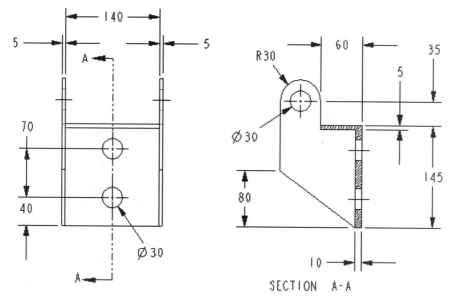

Figure 9 Dimensions (mm) of mounting bracket model

Defining Surface Pairs

When your Pro/E model is ready, launch MECHANICA with

>   *Applications > Mechanica*

Accept the default model type (3D). Select the *New Shell Pair* toolbar button or use

>   *Insert > Midsurface*

Try using **Auto Detect**. It won't work here, so we will have to create the shells manually. Select **New** and pick (using CTRL) the outer and inner surface of the vertical plate on the right side of the bracket, then middle click. The surfaces will highlight in red and yellow and a shell pair will appear under **Idealizations** in the model tree. Continue to pick pairs of parallel surfaces until all four pairs are selected. Then repaint your screen. Now select (in the MIDSURFACES menu)

> **Show > Select**

and pick on any surface. The paired surfaces will highlight. In the **MIDSURFACES** menu, select

> **Compress > Shells only**
> **ShowCompress**

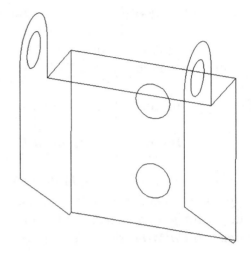

Figure 10 Compressed surfaces

The model will be replaced by the midplane surfaces as shown in Figure 10 highlighted in yellow. Leave the menu with **Done/Return**.

Check the **AutoGEM** menu. The option **Solid/Midsurface** is now checked. Since we don't have any solid parts to this model, change this to **Midsurface** alone.

Completing the Model

Assign the material AL2014 to the part:

> **Define Materials**
> {move AL2014 to the model list}
> **Assign > Part**

Click on the part, middle click, and accept the dialog.

Now apply constraints. As before, we will apply these to edges. Although this geometry is symmetric about the midplane, the loading will not be. Therefore, we can't use symmetry for this model. We will constrain the hole edges against translation only.

> **New Displacement Constraint**

Create a new constraint set called **[fixededges]**. Name

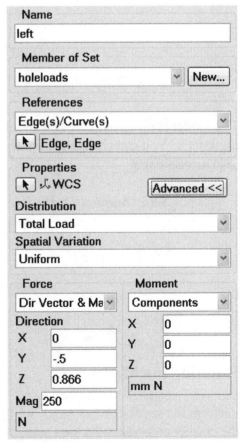

Figure 11 Specifying a force by magnitude and direction

the constraint *[holes]*. Set the **References** type to *Edges/Curves* and pick on the front edges of the two holes on the back plate of the bracket. Make sure you pick both halves of each circular curve. They will highlight in red. Middle click. Leave all translation degrees of freedom FIXED, and change the rotations to FREE. Accept the dialog.

Now apply the loads. Use the toolbar *New Force/Moment Load* button or select:

> *Insert > Force/Moment Load*

Create a new load set called *[holeloads]*. Name the first load *[right]*. Select the **References** type *Edges/Curves* and pick on the outside edges of the hole on the right vertical plate (both halves). They will highlight in red. Middle click. Select the *Advanced* button and confirm that *Total Load* and *Uniform* are the defaults. Set the X component to *100*. Accept the dialog.

Follow the same procedure for the hole in the other vertical plate. Name this load *[left]* (in load set **holeloads**). Select the edge of the hole on the left vertical plate. Select a **Total Load**, **Uniform** distribution. In the **Force** pull-down list, select **Dir Vector & Mag**. We want a force 30° below horizontal, so enter the vector components (**0, -0.5, 0.866**) in the X, Y, and Z directions, respectively. Enter a magnitude of **250**. See Figure 11. Accept the dialog.

Note that we applied the loads and constraints to edges, not surfaces (why?) and that we did not have to specify a shell property (thickness) - this is obtained from the Pro/E solid model.

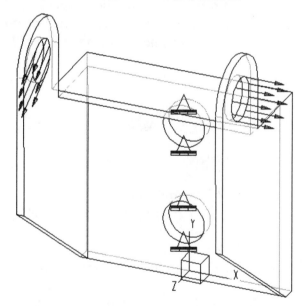

The model is now complete and should look like Figure 12. You can change the attachment of the load arrows using

> *Setup Simulation Display*

Figure 12 Mounting bracket model complete

Running the Model

Perform the usual analysis steps: set up and run a **QuickCheck** analysis. Make sure you select the correct constraint and load sets. AutoGEM will create something like 45 shell elements. The maximum Von Mises stress is about 22 MPa. *Edit* the analysis to run a **Multi-Pass Adaptive** analysis (5% convergence, maximum edge order 9). The multi-pass analysis does not converge on pass 9 - an indication that something is wrong. The maximum Von Mises stress has increased to 61 MPa which seems a little high (3 times the QuickCheck value).

Create some result windows to show the Von Mises stress and the deformation animation. The deformation is shown in Figure 13. Note the scale of the display. The maximum displacement is only about one-half millimeter. This looks fairly reasonable.

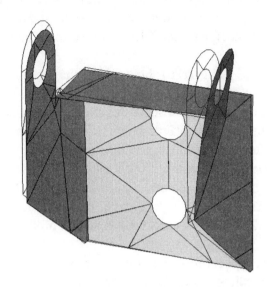

Now bring up the display of the Von Mises stress. It first appears as in Figure 14. Almost the entire model is shown in the lowest 2 or 3 fringe colors (below 20 MPa). There are two very small "hot spots" at the corners of the vertical plates. Show the location of the maximum Von Mises stress. It occurs at the top corner on the right plate. To see the rest of the stress distribution in the part, we need to redefine the stress levels assigned to the colors in the legend.

Figure 13 Deformation of the bracket

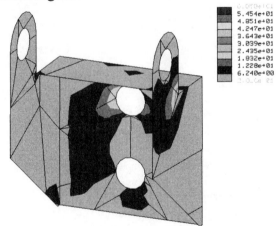

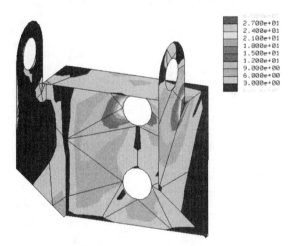

Figure 14 Von Mises stress fringe plot with default legend - not very useful here!

Figure 15 Von Mises stress fringe plot with edited legend - much better!

In the pull-down menu, select

Format > Legend

Change the values to read

Maximum	**27**
Minimum	**3**
Levels	**9**
Color Spectrum	**Mechanica Classic**

then select **OK**. Notice the combination of the minimum and maximum values and number of levels in the legend (3 x 9 = 27). This produces "nice" legend values that are easier to read. We have also put all the fringes in the lower half of the stress scale (note the maximum reported is over 60). Turn off shading with **View > Shade**. The "hot spots" are now much more visible, as well as the stress distribution around the mounting holes. See Figure 15 Feel free to experiment with other settings for the legend levels and fringe color spectrum.

Create windows to show the convergence history of the maximum Von Mises stress and the total strain energy. These are shown in Figure 16. The Von Mises stress is increasing steadily with each pass with no sign of converging at all, while the strain energy does seem to be converging. This behavior coupled with the stress contours indicates that there is a singularity at the sharp inside (concave) corner. The p-code method in MECHANICA is not able to converge on this type of geometry, as it will continue to try to increase the polynomial order indefinitely in order to catch the (theoretically) infinite stress at the corner. This means that any results (especially the stress) reported right at the corner (and in the immediate vicinity) must be taken with a large grain of salt - it cannot be trusted at all! This geometry is called a *re-entrant corner*.

The presence of re-entrant corners and their effect on convergence analysis is a major complication for the FEA analyst.

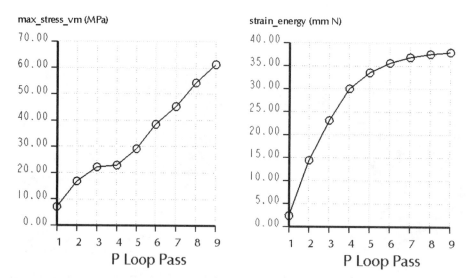

Figure 16 Convergence graphs of Von Mises stress (left) and strain energy (right)

Figure 17 below shows a closeup of the mesh created at the left re-entrant corner. To see this mesh, go to the AutoGEM menu and select **Create**. You can also select

Setup Simulation Display > Mesh

and turn on the option to shrink elements. AutoGEM creates several small elements at this re-entrant corner. Inexplicably, AutoGEM does not create these small elements at the right side - it probably should.

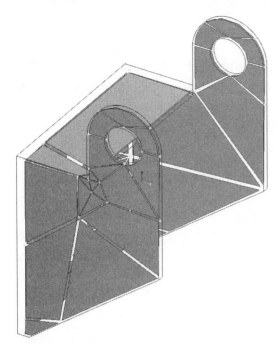

Figure 17 Default mesh showing elements at left re-entrant corner

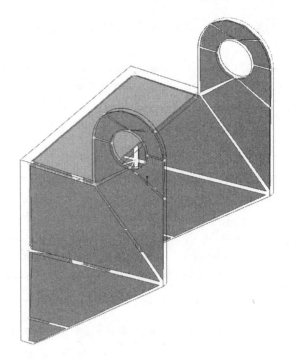

Figure 18 Mesh created with **Re-entrant Corner** option turned *OFF*

In independent mode, you can tell the program to ignore results on elements at a re-entrant corner (they are *excluded*) while it monitors convergence on the rest of the model. In fact, exluded elements can be automatically selected for you. Unfortunately, this cannot be done in integrated mode. You must therefore be on the lookout for this type of behavior, especially with shell models.

In the AutoGEM settings dialog, set the following options:

Turn *ON*   **Delete Mesh Points When Deleting Elements**
Turn *OFF*   **Re-entrant Corners**

The first option causes all traces of any previous mesh to be eliminated when a mesh is deleted. Otherwise, node points are left in the model that will be incorporated in the new mesh. Turning off the second option tells AutoGEM to ignore potential re-entrant corners and just produce a simple mesh. Return to the main AutoGEM menu and select *Create* again. This time there are only 33 elements. The left corner is shown in Figure 18. Notice the absence of small elements at the corner.

Rerun the MPA analysis to see what happens to the model results with this new mesh. What happens to the maximum stress? Convergence? If results have changed, explain why. If they haven't changed, explain that too!

Mixed Solids and Shells (Model #3)

Quite often, a solid part will contain regions of thin-walled material. The part could, of course, be modeled completely using solid elements. However, it will usually be more efficient to model any thin-walled features using shell elements, leaving the rest of the model as a solid. This is illustrated by the example in this section. We will also look at the *Bearing* load, and some new forms of boundary constraint.

The part we will model, a simple (large!) bell crank, is shown in Figure 19. The base feature for this part is a simple sweep, shown in Figure 20. The dimensions for the part (in millimeters) are shown in Figure 21. Start this new part called *[crank]* using the mm-N-s part template. Notice the orientation of the default coordinate system in Figure 19. The sweep can be created using a sketched trajectory on the TOP datum plane. The central of the three bosses at the corner of the sweep is located at the origin. After the sweep, create the three bosses and then coaxial holes (all *Thru All*).

When the model is completed, transfer into MECHANICA with

> *Applications > Mechanica*

Remember that the default model type is 3D, so we don't need to do anything about the model type.

Figure 19 Bell crank model

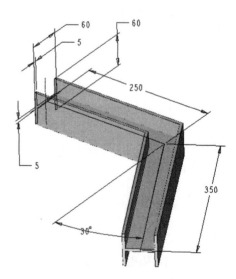

Figure 20 Base feature - Sweep

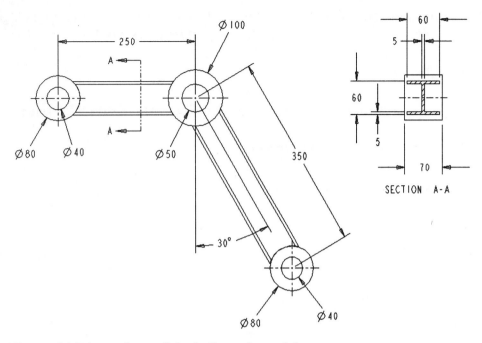

Figure 21 Dimensions of the bell crank model

Creating the Shells

The first thing to do is identify where we want shell elements. Select (or use the *New Shell Pair* button in the toolbar):

> *Insert > Midsurface*

Try to use the *Auto Detect* command. This will not work here - it only works if the thin-walled feature was created using the *Shell* command (and some others like *Rib*) in Pro/E. We will again have to create the shell surface pairs manually.

In the **MIDSURFACES** menu, select

> *New > Constant*

Pick on the upper and lower surfaces of the middle of the crank arm section (the cross piece in the H-section). It might help to be in Shaded display mode here. When the two surfaces are selected, middle click. One surface will be red and the other yellow. Continue picking pairs of surfaces. There are five pairs in all. Be careful that when picking the remaining pairs for the vertical sides of the crank arms, the first surface selected is the larger one on the outside of the arms. You must also pick both surfaces on the inside (above and below the cross piece).

When all five shell pairs are defined, select *Show* and pick on any of the surfaces to confirm that they are identified properly. Now select

> *Compress > Shells Only*

This will display a wireframe with red representing solids, green representing original geometry, and yellow highlight for the shell midsurface. In the **COMPRESS MDL** menu, select

> ### Show Paired

This will display just the shell surfaces (Figure 22). Finally, (**IMPORTANT!**) select

> ### Shells and Solids > ShowCompress

and you will see the complete model in Figure 23.

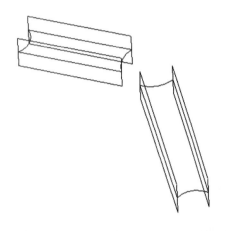

Figure 22 Shells using ***Show Paired*** option

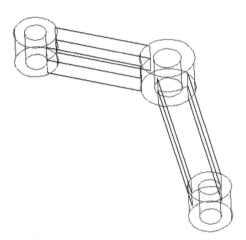

Figure 23 Model displayed using ***Shells and Solids > ShowCompress***

Return to the pull-down AutoGEM menu to make sure that the ***Solid/Midsurface*** option has been checked - we need both for this model.

We can now proceed to define the rest of the model.

Defining the Constraints

For the constraints, we want to restrict motion of the central boss and the one at the end of the long crank arm. We will constraint each hole surface against radial and axial motion, but allow rotation around the hole axes. To do this, it is most convenient to have a cylindrical coordinate system centered on each hole, with the Z-axis of the system lined up with the hole axis.

To create the coordinate systems, start with (in the pull-down menu) (or use the ***Datum Coordinate System*** toolbar button)

> ### Insert > Model Datum > Coordinate System

In the **Type** pull-down list[2], select the option *Cylindrical*.

Now (using CTRL) pick the datum planes **TOP**, **RIGHT**, and **FRONT**. A cylindrical system is placed at the intersection of these three planes. Use the *Orientation* tab menu to set the *Z-axis* so that it lines up with the hole, and the *Theta=0* axis is parallel to the default X-axis. This new system is **CS0** (check the model tree).

We want to create another coordinate system (**CS1**) at the hole on the long arm of the crank. Once again, the *Z-axis* should line up with the hole. Pick the hole axis as the first reference. The TOP datum plane is the second placement reference for the origin. Use the RIGHT datum plane to set the orientation for the *Theta=0* direction.

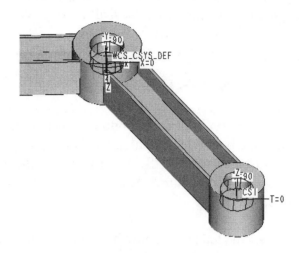

Figure 24 Cylindrical coordinate system defined on hole axes

The two cylindrical systems are shown in Figure 24. Now we can define the constraints on the hole surfaces. In the pull-down menu, select (or use the *New Displacement Constraint* button)

Insert > Displacement Constraint

Name the constraint *[hole1]* (in ConstraintSet1). Select the **Surface** button and pick on the surfaces of the hole at the origin. Middle click. Select the **Coordinate System** button and pick **CS0** (you might find the model tree useful here) as the reference coordinate system. Now set the constraints on R (**FIXED**), Theta (**FREE**), and Z (**FIXED**). We will allow the boss/hole to rotate, but cannot change size or position. Recall that the rotation constraints will have no effect on these surfaces since they will be used for solid elements. Accept the dialog with *OK*.

Repeat this procedure for the constraint on the other hole. Call it *[hole2]* (also in ConstraintSet1) and select the reference coordinate system **CS1**. Accept the dialog.

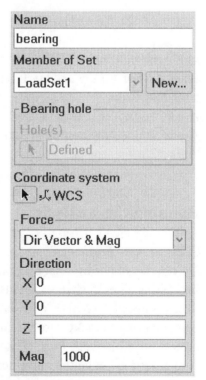

Figure 25 Defining a bearing load

Defining a Bearing Load

We will apply a bearing load on the hole at the end of the shorter crank arm. A bearing load has a resultant force in a

---

[2] We wonder how long before the PTC folks correct the spelling of "Cartesian" here.

specified direction. The actual force is applied normal to the bearing surface in a non-uniform distribution, more or less to model what would happen if a solid shaft were placed in the hole and a lateral force applied on the shaft in the direction specified.

In the pull-down menu, select (or use the ***Bearing Load*** button)

> ***Insert > Bearing Load***

This brings up the dialog window shown in Figure 25. Name the load ***[bearing]*** (in LoadSet1). Select the button under Hole(s) and click on the hole surface. Middle click. Note the selected coordinate system (WCS). We can specify the direction in three ways. In the Force pull-down list, select **Dir Vector & Mag**. Enter the data shown in Figure 25. Finally, enter a magnitude of **1000**.

Preview the load to see how it is being applied. This gives you some idea of how the load is distributed. Accept the dialog.

Defining the Material

Select the toolbar button to

> ***Define Materials***

Move the material **STEEL** into the model list. Then ***Assign > Part*** and click on the part. Middle click and ***Close***.

The model is now complete and should look like Figure 26.

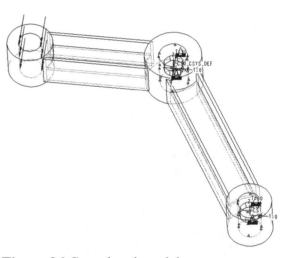

Figure 26 Completed model

Running the Analysis

Create a *New Static* analysis called *[crank]*. As usual, start with a **Quick Check**. Verify your ***Run Settings*** before selecting *Start*. Open the ***Study Status*** window. AutoGEM creates something like 35 shell elements and 224 solid elements. The maximum indicated Von Mises stress is about 16 MPa. You should have a look at the results of the Quick Check (especially the deformation) to make sure the model is doing what you expect it to. Notice the rotation of the bosses. If all is well, ***Edit*** the analysis to set up a **Multi-Pass Adaptive** analysis with 10% convergence. Leave the maximum edge order at the default, 6. Run the MPA analysis.

 * The MPA run will take a bit longer than usual! (3 - 10 minutes depending on your hardware)*

The MPA analysis does not converge on pass 6. The Von Mises stress has increased to 30 MPa. This indicates either that we should modify the mesh parameters to have AutoGEM create more elements, or we have something like a singularity condition.

Reviewing the Results

Create the usual result windows to show a fringe plot of the Von Mises stress, a deformation animation, and the convergence of the Von Mises stress and strain energy.

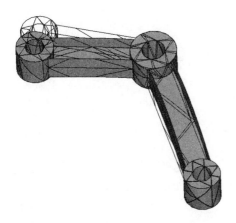

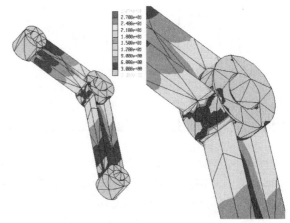

Figure 27 Deformation of bell crank

Figure 28 Von Mises stress in bell crank

The deformed shape is shown in Figure 27. Observe that the constraints are doing what we want - the bosses can rotate around their axes, but cannot move axially or radially. You might have to zoom in on the end boss to see the rotation.

The Von Mises stress is shown in Figure 28. Find out where the maximum stress is located. This may be the location of our convergence problem.

Finally, the convergence graphs are shown in Figure 29. There is clearly a problem with this model in regards to the Von Mises stress. The strain energy has converged (more or less) after the 5th pass, but the stress keeps rising. This is a sign of a singularity condition and it is difficult, therefore, to say anything conclusive about these results[3].

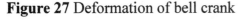

[3] The author debated a long time about including this exercise in the lesson since it may leave the reader with a rather low opinion of this type of model. It was decided to include it specifically because it displays some of the diabolical behavior that must be dealt with by the FEA analyst! FEA is not always as straight forward as many people think, and many models will give you problems with convergence. Notwithstanding the performance of this model, mixed solid/shell models are very useful and often necessary.

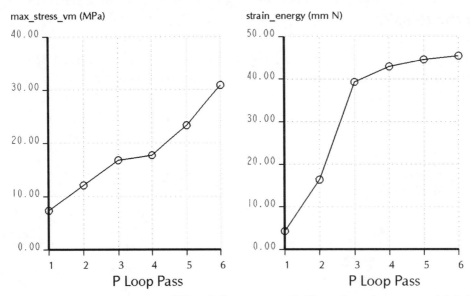

Figure 29 Convergence of Von Mises stress (left) and strain energy (right) for the bell crank model

To compare with a solid model of the same part, go to the **AutoGEM** menu and select the **Solid** option. Launch the **Create p-mesh** command from the top toolbar; AutoGEM will create almost 1000 solid elements. Observe the shape of the elements formed in the thin wall of the section. Despite the ruggedness of p-elements, we might expect the required edge orders to be high. With this many elements, then, we would expect a significantly longer run time than we saw before. Try it! You may be surprised at the results.

Summary

This lesson has introduced you to the main tools for using shell elements in general 3D models. Shells are created by compressing thin features formed from parallel straight or curved surfaces. For certain models, these surface pairs can be selected automatically. Otherwise, you can select paired surfaces manually. In either case, the shell thickness is read from the actual Pro/E model, rather than specified as a separate shell property (as in axisymmetric shells).

One of the main problems you will face with shell models is re-entrant corners. These will dominate the convergence behavior. Even worse, they leave you guessing at the correct stress levels in precisely the areas where they are highest. It might be possible to modify the mesh somewhat near these singularities using the mesh control methods discussed in a previous lesson (datum curves, edge controls, datum points, AutoGEM limits, etc.). Another possible way around the singularity problem is to monitor convergence at points removed from the singularity location, perhaps by seeding the geometry with datum points. Measured quantities can be set up at these points and monitored for convergence, rather than looking at the model maximum.

We also saw how a bearing load can be created and some variations on specifying constraints that allow rotation around an axis.

Despite the somewhat dubious performance of our third model, shell elements can be very beneficial in models with thin-walled features. They can drastically reduce the number of elements in the model and the computation time to get a solution. As always, we must be very careful about interpreting these results.

In the next lesson, we will look at another of the idealizations covered in this book. This idealization is used in the treatment of beams and frames.

Questions for Review

1.   When will *Auto Detect* locate surface pairs of thin-walled features? (For what Pro/E features and geometry will it *not* work?)

2.   The *Shell* command in Pro/E can be used to create shells with different thicknesses on different surfaces. How does this affect *Auto Detect*?

3.   Suppose you have a model with a number of thin-walled features (say 1.0mm thick) and also have several parallel surfaces slightly thicker (say 10 mm thick) that you want to treat as a solid. How does *Auto Detect* behave in this situation?

4.   Can you delete a single shell pair from a set found using *Auto Detect*? What happens if you still try to execute the model?

5.   What happens if, when selecting surface pairs manually, you accidentally pick two surfaces that intersect?

6.   Can you manually pick concentric cylindrical surfaces as members of a surface pair?

7.   How do you specify the material for a model composed entirely of shell elements?

8.   What type and shape of elements does AutoGEM create for general 3D shells?

9.   Is it possible to create a model with two shells that pass through each other?

10. What are the restrictions on applying loads and constraints to shell models?

11. How do you specify the thickness of a shell model? Does this have to be the same everywhere in the model?

12. An end view of a small portion of a solid part is shown in the figure. Each of the three thin "spokes" extend into the page. We want to model each spoke as a shell. Do you anticipate problems? If so, what?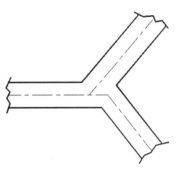

13. Suppose you are making a model of a thin-walled plastic part, which must have draft on all vertical surfaces. How will this complicate your model?

14. What are the symptoms of the presence of a singularity in the FEA model? How will this complicate the analysis of the results of your model?

15. Normally, if a model will not converge with the maximum edge order (9), the solution is to modify the mesh to produce more, smaller elements (mesh refinement). Will this strategy work in the presence of a singularity? Why?

16. If the maximum Von Mises stress result is not suitable for monitoring convergence in a model with a singularity, what other measures could you use to monitor convergence? (See also exercise #4.)

Exercises

1. A U-shaped steel beam is cantilevered out from a wall and carries the loads shown. The
    beam is a C-channel with dimensions given on the right. Find the magnitude and location
    of the maximum Von Mises stress, and the maximum displacement.

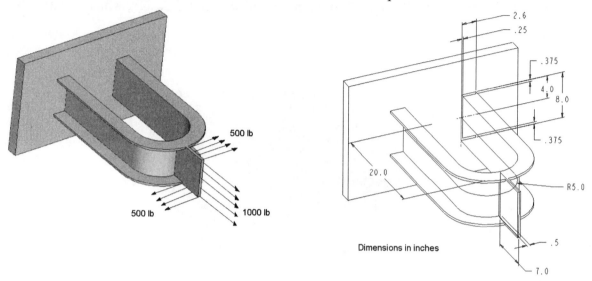

2. Make a solid model of the centrifuge used in Chapter 6, somewhat modified as shown in
    the figure below. (HINT: in Pro/E use a *Thin, Revolved Solid*). Bring this into Mechanica
    and model it using shells. Apply the same centrifugal load (50 rev/second). Can you use
    Auto Detect? The material is **AL2014**. How can you use symmetry to reduce the model
    size? Set up an optimization to find the minimum mass without exceeding a stress of 25.0
    MPa. The design variables are the thickness of the shells (*thick5* and *thick8* in the previous
    model) and the radial location of the inner vertical support. How much was the mass
    reduction from the initial design? Show this in a graph (mass vs optimization iteration).

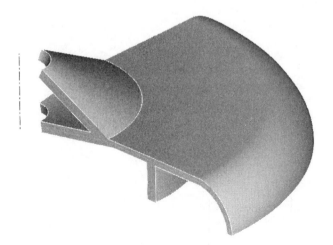

3.   Continue the optimization of the previous question adding the following factors: The
      exterior of the centrifuge is an evacuated chamber (to reduce air friction) while the inside of
      the centrifuge is pressurized to 30 kPa as shown in the figure below. In addition, we cannot
      allow the maximum horizontal or vertical deflection anywhere to exceed 0.02 mm. Find the
      thickness of the shells and the location of the vertical support that will result in the
      minimum inertia around the axis of rotation. For the optimized design, which constraints
      are active?

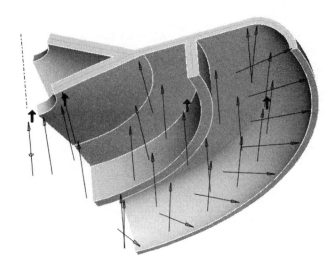

4.   Consider the bracket model again (model #2). We saw that the model's maximum Von
      Mises stress was an unsuitable measure for monitoring convergence when the model
      contains a re-entrant corner or other singularity. Create the datum curve (radius = 10mm)
      and point shown on the part below. Find out how you can define a measure to monitor the
      Von Mises stress at this point during an MPA analysis. Then, monitor this measure to
      determine convergence. Compare these results to those obtained during the lesson and
      comment on the validity of this procedure.

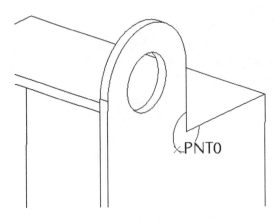

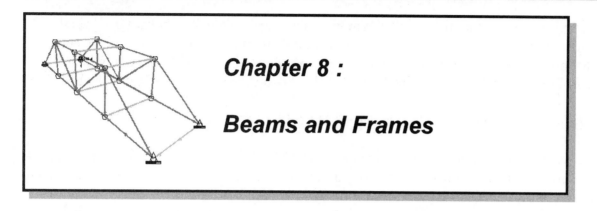

Chapter 8 :

Beams and Frames

Synopsis

Beam elements in 1D, 2D, and 3D problems; beam coordinate systems, sections, and orientation; distributed loads; beam releases; shear and bending moment diagrams; 2D and 3D frames; gravity load; displacement constraint

Overview of this Lesson

Beams are fundamental structural elements. In MECHANICA they are treated as idealizations. Beams are one-dimensional elements that can be either straight or curved. The element length must be much greater than the cross section dimensions (a rule of thumb is 10 times). In integrated mode, the cross section must be constant along the beam. A model might be composed entirely of beam elements, or they can be used in conjunction with other model entities (solids and/or shells).

This lesson introduces the main concepts required to model isolated continuous beams and beam elements as components of frames. The difficult subject of beam coordinate systems is introduced with sufficient depth to handle problems with simple symmetric beam cross sections. Beam orientation in 3D requires a solid understanding of these coordinate systems.

Four example problems are used to illustrate the MECHANICA commands. The first is a simple cantilever beam (a diving board) with a tip load. New results windows are created to show the shear force and bending moment diagrams for the beam. The second example is a more complicated continuous indeterminate beam. This introduces distributed loads and the use of beam releases. The final example is in two parts: a simple 2D frame and a 3D frame. These illustrate more ideas in beam orientation and gravity load. In the final model, the loading caused by a specified displacement of a constraint is used to model the settling foundation beneath one corner of the 3D frame.

Beam Coordinate Systems

One of the potentially confusing issues arising in the use of beam elements is their orientation with respect to the World Coordinate System, WCS. This is particularly true for curved beams, and beams whose cross sections are asymmetrical about their centroid in at least one lateral direction (such as channels and angles) and/or are offset from the underlying geometric curves. For the most general case, this orientation is described/defined using up to three coordinate systems. For this lesson, we only need to worry about two of these - the BACS and the BSCS[1].

The Beam Action Coordinate System BACS

On the screen, beam elements are represented as light blue lines. Beams are associated either with geometry curves or connecting two (or more) points defined in the WCS. The curves and/or points can be created either in Pro/E (as datums) or in MECHANICA (as simulation features). We will deal only with straight beams in this lesson. For these, the underlying curve or points will define the X-axis of the beam's BACS (see Figure 1). The beam's local Y- and Z-axes are perpendicular to the beam. The orientation (relative to the WCS), ie. rotation of the beam around its local X-axis, is defined by specifying the direction that the BACS Y-axis is pointing. This direction can be specified in a number of ways: an axis direction, an edge, a point, or giving vector components in the WCS. Some simple examples showing the specification of the BACS Y-Axis using vector components are shown in Figure 2. All the properties of each beam element are set up in the **Beam Definition** window. This includes the geometric references, material, orientation of the Y-axis, section shape, and so on. Several beams with the same properties can be created simultaneously.

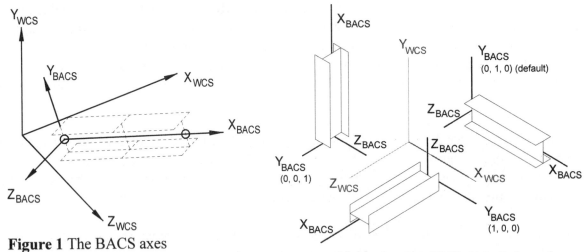

Figure 1 The BACS axes

Figure 2 Illustrating BACS Y-Axis Orientation

The Beam Shape Coordinate System BSCS

The beam cross sectional shape and position are defined relative to the BSCS. The standard cross sections built in to MECHANICA are shown in Figure 3. You can also create your own section shapes using Sketcher. The shape is defined in the BSCS YZ plane. For most standard shapes, the origin of the BSCS coincides with the centroid of the section. The X-axis of the BSCS

---

[1] The third system is the BCPCS (Beam Centroidal Principal Coordinate System). See the on-line help for further information on this system.

(coming out of the page in Figure 3) is always parallel to the X-axis of the BACS, that is, along the beam. The BSCS origin (or its shear center) is defined by offsets DY and DZ measured from the origin of the BACS. See Figure 4. The orientation of the BSCS is determined by the angle theta specified in the Beam Orientation property window.

The BSCS is parallel to the BACS if theta is zero. In addition, if the offsets DY and DZ are zero, then the BSCS coincides with the BACS. This will be the case in all the examples used in this lesson.

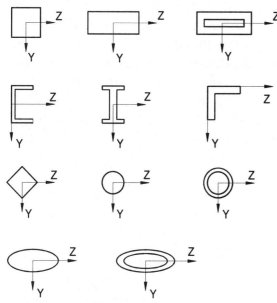

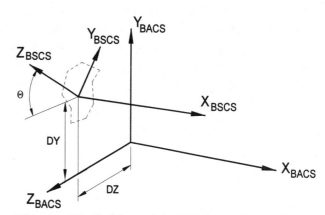

Figure 4 Definition of the BSCS axes relative to BACS. The frames coincide when theta, DY, and DZ are all 0

Figure 3 Standard Beam Section Shapes defined in the BSCS axes

When a beam cross section and orientation are specified, the combined properties will appear as an icon in true scale at four locations along the beam element. This serves as a visual cue to the shape, size, and orientation of the beam. Some examples are shown in Figure 5. These icons also indicate the directions of the BSCS Y- and Z-axes. The Y-axis is an open V (the Y axis is upward in Figure 5), while the tip of the Z-axis is an open arrowhead.

Now, this all seems pretty complicated, and indeed it is. Fortunately, in most cases, the MECHANICA defaults are exactly what is required and you can do a lot of modeling knowing only the bare essentials.

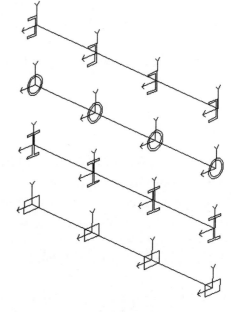

Figure 5 Beam elements with section icons

Example #1 - Basic Concepts

Our first example is a simple beam similar to a diving board. This will introduce some of the concepts involved in using beam elements.

The Model

The model is shown in Figure 6. The cross section is a hollow rectangle, 24 inches wide and 2 inches high, with a wall thickness of 0.125 inch. The beam is cantilevered out from the wall, and rests on a simple support 10 feet from the wall, with an overhang of 6 feet. The material is aluminum and a downward vertical load of 200 lb is applied at the tip. This is a static load, and we will calculate the shear force and bending moment and static deflection, as if someone was just standing at the tip, not bouncing up and down. Note that this problem is statically indeterminate - that is, the simple methods of statics cannot be applied because of an extra unknown in the reactions. Nonetheless, analytical methods[2] could be used to solve for such things as the bending moment along the beam and the deflection at the tip.

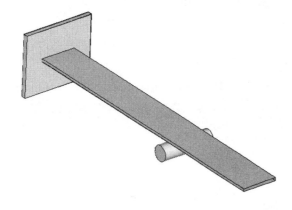

Figure 6 The diving board model - a simple indeterminate beam

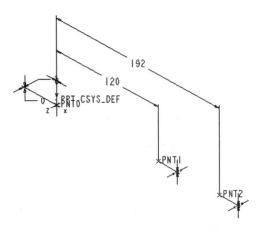

Figure 7 Datum points created in Pro/E

Geometry

Create a new part in Pro/E called *[divingboard]*. You can use the default part template, but change your units to in-pound-sec (IPS). The beam will be created along the X-axis of the default coordinate system. Since we are going to use an idealization of the beam, all you need to do here is create three datum points (perhaps using the *Offset Coordinate System* toolbar button on the *Datum Point* toolbar flyout). Create the three points at the locations (X, Y, Z) = (*0, 0, 0*), (*120, 0, 0*), and (*192, 0, 0*). See Figure 7.

With the datum points created, transfer into MECHANICA with

> *Applications > Mechanica*

---

[2] See the *singularity method* discussed in most mechanics of materials textbooks.

Keep the model type default (3D).

Beam Elements

We will create two beam elements directly from the points. Select the following (or use the toolbar button for **New Beams**):

> *Insert > Beam*

The **Beam Definition** window appears. It will look like Figure 9 when we are finished. Accept the default name **Beam1**. The geometry reference we want is the default (Point-Point). Select the button under References and pick the point at the origin (**PNT0**) and the point that will be under the support (**PNT1**). Beside the **Material** pull-down list (currently empty), select *More*. Move the material **AL2014** over to the model list then select *OK*. The default Y-direction is (0,1,0), that is, parallel to the WCS Y-axis - exactly what we need. Beside the **Section** pull down list, select *More > New*. This brings up the **Beam Section Definition** window shown in Figure 8. Name the section *[hrectangle]* and in the **Type** pull-down list select **Hollow Rect**. Enter the dimensions beside the sketch of the cross section as shown in Figure 8. Accept the dialog. Back in the **Beam Definition** window, notice that **Orientation** is set to **(none)**. This refers to the BSCS offset and angle settings (DX, DY, and Theta) shown in Figure 4. By default, these are all zero and the BSCS aligns with the BACS. In the **Beam Definition** window, select *OK*. The first beam element appears. The cross section shape icon is in true scale.

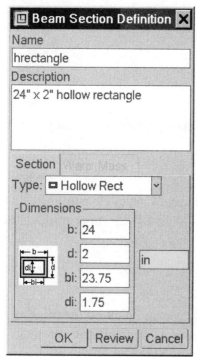

Figure 8 Specifying the section properties

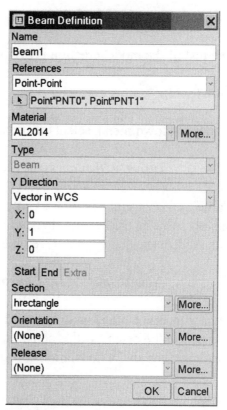

Figure 9 Defining a beam element

Now create the second element. In the toolbar, select *New Beam*. The definition window opens again with all the data fields already entered. The default name is **Beam2**. All you have to do here is select the new points (**PNT1** and **PNT2**) to define the References. Accept the dialog. Both beam elements are now defined, see Figure 10. Notice the section Y and Z directions on the section icons.

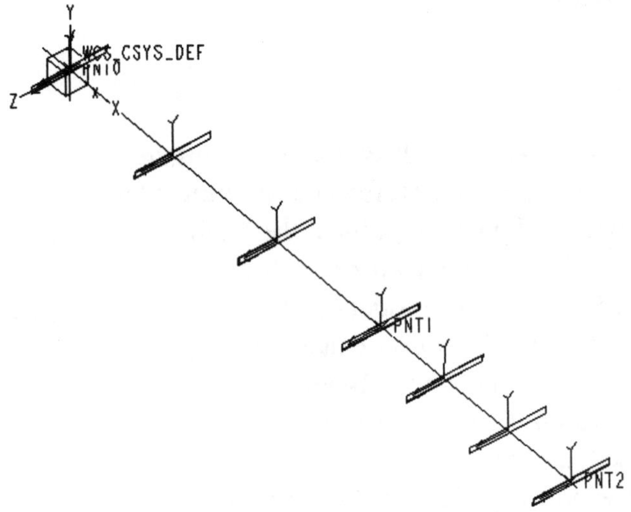

Figure 10 Model with beam elements defined

Completing the Model

The rest of the model creation is pretty routine.

Constraints

In thepull-down menu, select (or use the *New Displacement Constraint* toolbar button)

> *Insert > Displacement Constraint*

Name the constraint *[fixed]* (in **ConstraintSet1**). Set the **References** type to *Points*, and click on the left point on the beam; middle click. Leave all the constraints **FIXED** for this cantilevered end. Accept the dialog and move on to the other constraint:

> *New Displacement Constraint*

Name this one *[roller]* (also in **ConstraintSet1**). Set the *Points* type and pick on the middle support on the beam. At this constraint we want to simulate a roller, so free the Z rotation constraint and the X translation constraint. Accept the dialog.

Loads

We'll apply a single point load on the tip. Select (or use the *New Force/Moment Load* button):

Insert > Force/Moment Load

Name the load *[download]* (in **LoadSet1**). Set **References** to *Points*, and pick the point at the tip; middle click. Enter a Y component of *-200*. Accept the dialog. Use *Setup Simulation Display* to change the arrow setting to **Tails Touching**. In the same window, turn on **Display Names** and turn off **Load Value**. This completes the model, which should now appear as Figure 11.

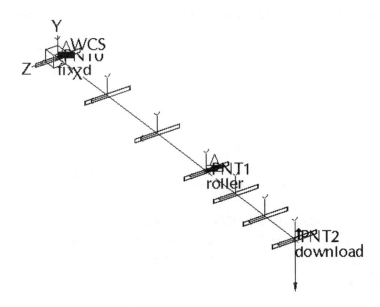

Figure 11 Diving board model completely defined

Analysis and Results

Performing the analysis involves the usual steps. Open the *Analyses/Studies* window and select

File > New Static

Enter a name *[divboard]*. Make sure constraint and load sets are selected. Set a convergence method *QuickCheck*. Select the **Output** tab and change the **Plotting Grid** to **10** - this will put more points along our graphs a bit later. Then continue the *Run Settings*. Set up the usual locations for temporary and output files. Then accept the dialog and *Start*.

Open the *Study Status* window and check for errors. Assuming there are none, change to a *Multi-Pass Adaptive*. Set a 1% convergence criterion. You can leave the maximum polynomial order at *6*. Re-run the analysis. Delete the existing output files and open the *Study Status* window. The run converges on the 2nd pass, maximum edge order 4, with zero error. The maximum bending stress **max_beam_bending** is 2,670 psi, and the maximum displacement is -0.998 inch in the Y direction.

Now on to the result windows.

Deformation and Bending Stress

We will create the usual result windows for stress and deformation, with one slight difference. Name the first window *[bending]*. Get the output directory **divboard**. Select a **Fringe** plot and select *Quantity(Stress , Beam Bending)*. Under the **Display Options** tab, check the box beside **Deformed** and **Overlay Undeformed**.

Copy this window to another called *[deform]* and change the definition to produce a displacement animation.

When you *Show* these two windows, you have to be careful about the view orientation - check the XYZ coordinate triad. You are probably looking at the beam in the Pro/E default direction. Use the *Saved*

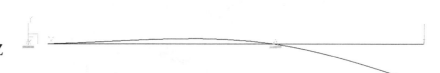

Figure 12 Deformed shape of the diving board

View List toolbar button to select the **FRONT** view. The bending stress figure is not reproduced here. It shows that the maximum bending stress occurs at the roller support. The deformed shape of the diving board is shown in Figure 12. Note that the slope at the left end is zero, as it should be for a cantilevered support.

Shear Force and Moment Diagrams

Now for some new forms of result windows: the shear force and bending moment diagrams. *All forces and moments are computed relative to the BACS.* Copy one of the existing windows to a new one called *[sfbm]*. Change the display type to *Graph*, and the **Quantity** to *Shear and Moment*. Deselect all options except *Vy* and *Mz* as shown in Figure 13. Leave the graph location set to **Beams**, then pick the *Select* button and, starting at the left end, click on each beam element. As you pick elements (they will highlight in red), follow the messages at the top of the window. Do not worry that the section icon seems to be in the wrong direction. When you have picked both elements, middle click. Note that the highlighted element end point will be on the left end of the graph.

Select *OK and Show*. We get the combined shear and bending moment diagrams for the two beam elements (note that the horizontal scales are different.) shown in Figure 14.

IMPORTANT POINT: From your knowledge of simple beam theory, what sign convention does MECHANICA use for shear and bending moment?

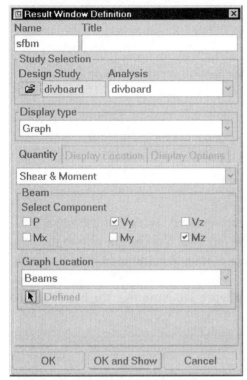

Figure 13 Result window definition for shear and bending moment diagrams

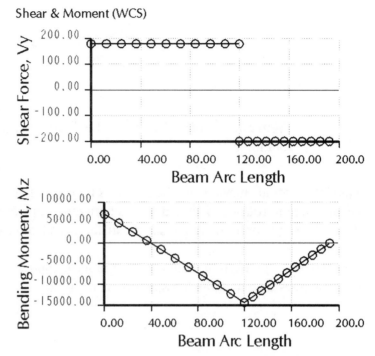

Figure 14 Shear force and bending moment diagrams

As a check on these results, the shear force should be constant on each element (and 200 on the right element). The bending moment should (at least) vary linearly on the right element, be zero at the right end, and non-zero at the left end. Qualitatively, this diagram looks fine.

As an exercise, find out what happens to these graphs if we reverse the direction of the BACS Y-axis.

Save the result window definition (rwd) file, since we can (partially) use it again later.

Changing the Constraint

We'll change the constraint on the left end to a pinned joint instead of cantilever. Find the constraint **fixed** in the model tree. Select it, and in the RMB pop-up menu, select

Edit Definition

Change the Z rotation constraint from fixed to **FREE**. Run the Multi-Pass Adaptive analysis again. The maximum displacement in the Y direction is now -1.17 inch, so it has increased a bit from the previous case - the diving board is a bit less stiff. The maximum bending stress is the same (Why?). Can you explain the value of the total strain energy?

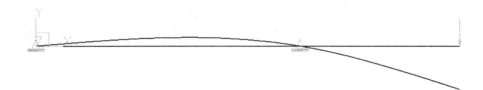

Figure 15 Deformed shape of the diving board with pinned left end (Z rotation FREE)

The deformation change from the previous case is not very pronounced. With a pinned end, the beam is free to rotate and should have a non-zero slope at the left end. You might like to increase the scale for the deformation to see the differences from the previous constraint case a bit clearer.

The new shear and bending moment diagrams are shown in Figure 16. Note that the bending moment goes to zero at the left support, as it should.

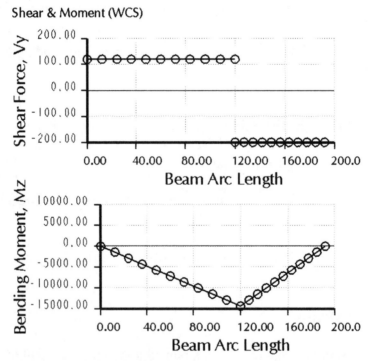

Figure 16 Shear force and bending moment diagrams for beam with pinned left end

Now on to the second example, which is a bit more complicated.

Example #2 - Distributed Loads, Beam Releases

This example is a bit more complicated and will introduce the use of distributed loads and beam releases.

The Model

A drawing of the model is shown in Figure 17. We will use SI units: lengths in meters, force in Newtons. The steel beam cross section is an I-beam. To accommodate the distributed loads, and to provide sites for beam releases in the second part of the example, the model is divided into 4 beam elements. The origin of the WCS XY system is at the left end.

IMPORTANT NOTE: This beam violates the MECHANICA guidelines for use of beam elements. This guideline is that the *ratio of a beam element length to its largest cross section dimension (its aspect ratio) should be greater than 10:1*, that is, the beam element should be long and slender. This is a normal assumption even for simple beam theory. In short, stubby beam elements, shear takes on an important role not accounted for in long slender beams. We are using this beam here strictly for demonstration purposes.

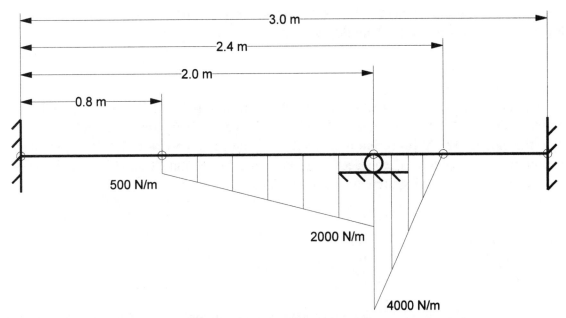

Figure 17 Indeterminate beam with distributed loads (dimensions in meters)

Beam Geometry

Start a new part in Pro/E called *[cbeam]*. Use the default part template, but change your units to the MKS system (meter - kilogram - second). The beam will lie along the X-axis of the default coordinate system. Create 5 datum points at the X locations (0, 0.8, 2.0, 2.4, and 3.0). These will be numbered PNT0 through PNT4.

In MECHANICA, distributed loads can only be applied to curves (not directly to beam elements). Therefore, we must create a couple of datum curves (these could also be created as simulation features in MECHANICA). These are between points PNT1 and PNT2, and between PNT2 and PNT3. See Figure 18. Use the **Datum Curve** toolbar button, and use the **Thru Points**, **Single Points** options.

With the Pro/E model completed, select

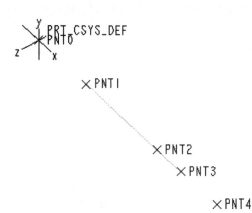

Applications > Mechanica

Note the force unit is Newtons. Accept the default model type (3D).

Figure 18 Pro/E model of beam (datum points plus two datum curves)

Beam Elements

We will create four beam elements. The end two use the datum points, the middle two use the datum curves. Select (or use the **New Beam** toolbar shortcut)

Insert > Beam

The first beam element name is **beam1**. The **References** type is **Point-Point** (the default). Pick the points **PNT0** and **PNT1**. Select the **More** button in the **Material** area. Select **STEEL** and add it to the model. The **Y Direction** is defined using the default **Vector** (0,1,0). Select the **More** button beside Section. Then select **New** and name the section *[ibeam]*. Select **Type(I-Beam)** and enter the following dimensions (notice that these are in meters):

flange width	b	*0.1*
flange thickness	t	*0.015*
web height	di	*0.10*
web thickness	tw	*0.01*

Select the **Review** button to see all section properties listed in the Browser (where you can easily get hard copy).

Accept all the dialog windows. The section icons appear on the beam. You might like to spin the view to see these clearly.

Select **New Beam** again. The next element is named **beam2**. Change the **References** type to **Edge/Curve**; pick the select button and click on the datum curve between PNT1 and PNT2. It highlights in red, and a magenta direction arrow is shown giving the direction of the BACS X-axis. If you click on the arrow, you can reverse its direction. Leave it pointing in the positive WCS X direction. Middle click. The **Y Direction** uses the default direction (0, 1, 0). The rest of the data is already filled in. Accept the beam element.

Create the third element, **beam3**, on the datum curve between PNT2 and PNT3 in the same way as beam2. Finally, create the fourth beam element, **beam4**, using Point-Point and picking the last two datum points PNT3 and PNT4. All the elements should appear as Figure 19.

The screen is going to get a bit cluttered up, so we can turn off the display of the beam sections and direction arrows. Select (in the top pull-down menu)

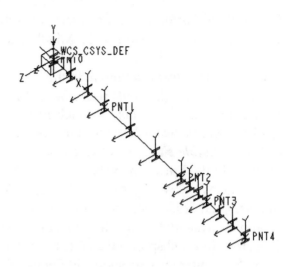

> *Setup Simulation Display*
> *Modeling Entities*

and deselect the option **Beam Sections**. Accept the dialog. All we see is straight lines representing the elements, without the section display.

Open the *Idealizations* entry in the model tree to see the beam elements listed there.

Figure 19 Beam elements created

Completing the Model

Constraints

The beam left and right ends are fixed (cantilever) and the middle support is a roller. These are all point constraints, with the difference being whether we allow rotation about the WCS Z axis. In the right toolbar, select

> *New Displacement Constraint*

Name the constraint *[fixed]* (in ConstraintSet1). Change the **References** type to *Points*. Select the far left and right points and middle click. Leave all the degrees of freedom as fixed. *Accept* the dialog.

Now for the point in the middle. Create a final displacement constraint here called *[midspan]* (in ConstraintSet1). Select on the middle point (PNT2). This is the roller support so free the X translation and Z rotation.

Distributed Loads

MECHANICA can define distributed loads only on curves (which is why we made them!). The load is transferred to the beam element(s) created on that curve. The load distribution can be set up using either built in functions (linear, quadratic, cubic, quartic) or specially defined user functions. We have two simple linear distributed loads in our model (recall Figure 17). The load on the model is obtained as a combination between a shape function and a magnitude. We will see two variations for how these can be used.

In the pull-down menu, select (or use the *New Force/Moment Load* button)

> *Insert > Force/Moment Load*

Call the load *[linload1]* (in LoadSet1). Set the **References** type to *Curves*, and pick the datum curve between PNT1 and PNT2. Middle click.

Now pick the *Advanced* button. In the **Distribution** pull-down list pick *Force Per Unit Length* and in the next list pick *Function of Coordinates.* Just below this, pick the button to *List Available Functions* (watch for the pop-up hint). In the **Functions** window, select *New*. In the next window (Figure 20), enter a name *[linear1]* and description. Note the default coordinate system. Change the **Type** to *Table*. Pick *Add Row* and specify the table to contain 2 rows. Now enter the data shown in Figure 20 - compare this to the load we are creating as shown in Figure 17. Select *OK* (twice) to return to the Force/Moment window. So far, we have created the distribution shape, but not actually applied it. In the Force/Moment window, enter a magnitude multiplier of **1.0** for force in the Y direction. *Preview* the load to check the functional variation and direction. See Figure 22. Notice how the function shape (Figure 20) and load value (Figure 21) combine to create the desired load. If all is well, select *OK* and proceed to the next load. When you accept the definition, the load icon will change.

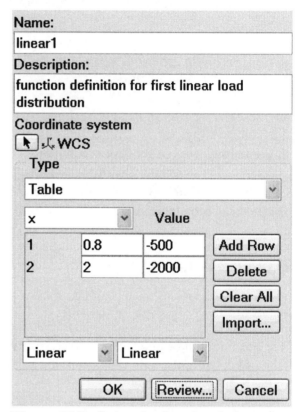

Figure 20 Defining the functional shape for a distributed load

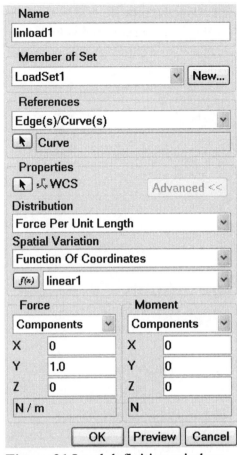

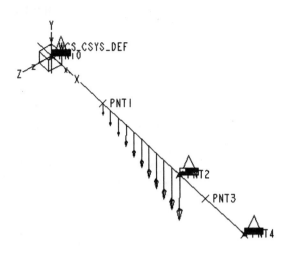

Figure 22 Preview of linearly distributed load on beam

Figure 21 Load definition window for distributed load #1

We'll create another distributed load on the next element along the beam using a somewhat different definition style. Start with the **New Force/Moment Load** toolbar button. Name this one *[linload2]* (in LoadSet1). Set the **References** type to **Edges/Curves** and pick the next datum curve. Once again use **Force Per Unit Length** and **Function of Coordinates**. Click the button to list available functions, then copy the definition of linear1 to a new definition called linear2. Edit the new function. Instead of directly entering load values, we will define only the shape here. In the table, enter the following data:

1	2.0	1
2	2.4	0

This sets up the desired shape of the function, with a unit magnitude. Accept the definition and back in the Force/Moment window enter a magnitude in the Y direction of -4000. To create the actual applied load, MECHANICA multiplies the magnitude times the shape function. This second method is perhaps more convenient since the load magnitude and direction are controlled directly from the Force/Moment window, instead of burrowing into the shape function.

Preview the load, and if it is satisfactory, select **OK**.

Change the display settings to **Tails Touching**. The display will show the relative size of the loads (but not the correct distribution). The complete model (without beam section display) is shown in Figure 23. The direction of the arrows between PNT1 and PNT2 is backwards because the sign of the force "component" for the first linear load is positive.

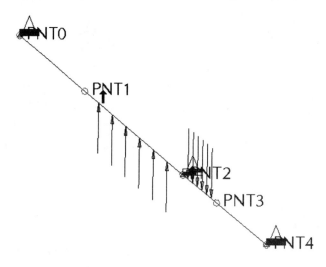

Figure 23 Completed beam model (section display turned off)

To review the two loads, select either in the model tree, then in the RMB pop-up select

Edit Definition > Preview

This will show the actual distribution shape and load direction. Repeat this for the other load.

Analysis and Results

In the pull-down menu, select

Analysis > Mechanica Analyses/Studies > File > New Static

Enter the name *[cbeam]*. The constraint and load sets (**ConstraintSet1** and **LoadSet1**) should be highlighted already. Select a *Quick Check* and using the **Output** tab, change the **Plotting Grid** to 10 - this will give us smoother curves in the result windows. Accept the analysis definition, check your *Run Settings*, and *Start* the analysis. Open the *Study Status* window and look for errors. Assuming there are none, set up a *Multi-Pass Analysis* with a 1% convergence. Leave the maximum edge order at 6. *Run* the new analysis. In the *Study Status* window you will see that convergence is obtained on pass 3, with not exactly a zero error (Why?). The maximum displacement in the Y direction is -2.35e-5 m (0.0235 mm), and the maximum bending stress is 1.34E6 (1.34 MPa).

Result Windows

We will create windows showing the deformation of the beam, and the shear force and bending moment diagrams.

> *Analysis > Results > Insert > Result Window > [deform]*

Get the output directory **cbeam**. Enter a window title "***Deformation***" and set up an animation of the displacement. The maximum deformation will look like Figure 24 (what is your view orientation and scale?). As usual, compare this with the anticipated result, and pay close attention to the constraints. The slope of the deformed beam looks like it is zero at each cantilevered end, and the deflection at the middle support is zero, as expected.

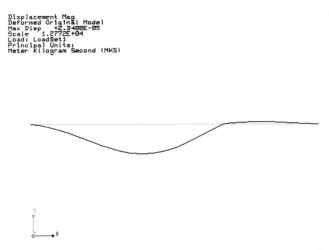

Displacement Mag
Deformed Original Model
Max Disp   +2.3488E-05
Scale   1.2772E+04
Load: LoadSet1
Principal Units:
Meter Kilogram Second (MKS)

Figure 24 Beam deformation (note scale factor!)

Now set up a new result windows *[sfbm]* to show the shear force and bending moment diagrams. Set up an appropriate title, select *Display Type(Graph)*, *Quantity(Shear and Moment)*, check *Vy* and *Mz*. Select *Graph Location(Beams)* and use the selection button to pick the four beam elements. Pay close attention to which end of the beam will be placed at the left side of the graphs. Select *OK andShow*. The result window graphs are shown in Figure 25, which has been reformatted for presentation here. We can look for things like continuity of bending moment along the beam, bending moment in the beam at each end, discontinuity in shear at the middle support, and so on.

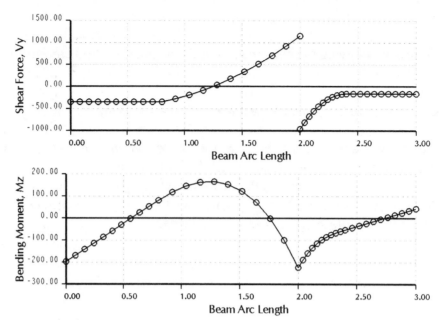

Figure 25 Shear force (top) and bending moment (bottom) diagrams for the I-beam model #2

Note the continuity (smoothness of curvature) of the deformed shape in Figure 24. In the shear and bending moments in Figure 25 note the non-zero bending moments at the connections between the first and second element (X = 0.8 m), and between the third and fourth element (X = 2.4 m). Note that you can easily obtain the shear force and bending moment data in a number of formats using *File > Export*.

Close the results windows.

Beam Releases

A beam release is used to change the type of connection between adjacent beam elements. For a normal (unreleased) connection, all six components of shear force and bending moment at the end of one element are carried through the connection to the next element. This results in continuity of these internal forces/moments along the beam (except at constraints or point loads). In this section, we will look at how we can interrupt this continuity using beam releases. In the present example, we will modify the model so that the beam is hinged at the second and fourth points. A hinge parallel to the Z axis of the element means that no bending moment M_Z can be transmitted through the connection. For static equilibrium of the beam on either side of the hinge, this requires that the bending moment be zero at the hinge.

Setting Releases

An important thing to remember here is that releases can only be applied directly to elements. This means only Point-Point beam elements. If the element is created using Edge/Curve (using a datum curve), it will not accept a release.

We will release two points. For the first one, open the model tree, select the first beam element. The element highlights in red. In the RMB pop-up, select *Edit Definition* and the **Beam Definition** window opens. The element starts at PNT0 and ends at PNT1. Near the bottom of the window select the **End** tab. Beside **Release**, select *More > New*. Enter a name *[releaseRZ]*. Select the *Rz* button at the bottom to free the rotation in the Z direction. Note that these are relative to the BACS (which in this model is parallel to the WCS). Accept the dialog and the beam definition. An icon close to the released point (at PNT1) is displayed to indicate the release. This consists of a small circle around an axis in the direction of the rotational release. We do not need to release the connecting element.

To make sure the release is in the model, open the model tree. Expand the **Idealizations > Beams** entries. Right click on **beam1** and select *Info > Simulation Object*. In the information window that opens, you should see **releaseRZ** listed for the end of the element. Close this window. You can also find the release in the Beam Definition window which can be brought up by right clicking on the element in the model tree and selecting *Edit Definition*.

Click on the fourth beam element (between PNT3 and PNT4). In the RMB pop-up, select *Edit Definition*. The end to release is at the **Start** of the beam. In the **Release** pull-down list, select **releaseRZ**. Accept the dialog and the beam definition. Observe the release icon on the element.

Results

With the new beam releases, rerun the multi-pass adaptive analysis. Delete the existing output files. Open the *Study Status* window and note that maximum bending stress is now 2.10 MPa and the maximum Y displacement is -3.81e-5m (-0.0381 mm). Compare these with the values obtained without the releases. The deformed shape of the beam with releases is shown in Figure 26. Note the abrupt changes in slope at the points of the beam releases.

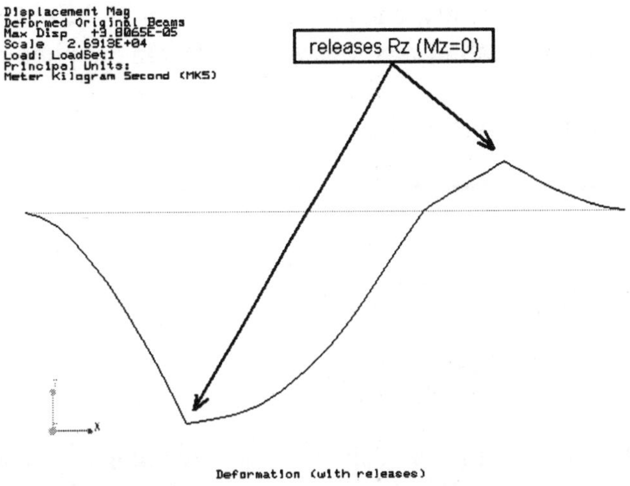

Figure 26 Deformation with beam releases

Now set up and display the shear and bending moment diagrams. See Figure 27. Observe that the shear Vy is non-zero and continuous across these released connections, and the moment has indeed become zero at the release locations.

Some other cases where beam releases will come in handy are in modeling trusses (no moment of any kind transmitted through a connection), an expansion joint (no axial load transmitted), or a connection like a dovetail (all forces and moments transmitted except shear in one direction).

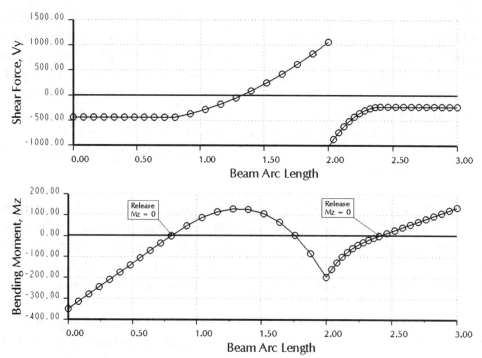

Figure 27 Shear force (top) and bending moment (bottom) for beam with releases

Note that in both of our beam examples, we have used loading only in the XY plane. This was for simplicity only, and is not a restriction in MECHANICA. We can apply loads in any direction (a situation called *skew bending*), including applied moments.

Close this model and take a break!

Example #3 - Frames

Model A - 2D Frame

The two previous models have been relatively simple one dimensional beams. Beams, of course, can be combined to formed complex 2D and 3D structures. In this section, we will investigate how to create frames based on the geometry shown in Figure 28. The material is steel and the beam section is a hollow circular pipe with an outside diameter of 3" and a wall thickness of 0.25". This shape will make specification of the beam orientation a bit easier. We will start off with a 2D frame, then move on to a full 3D frame.

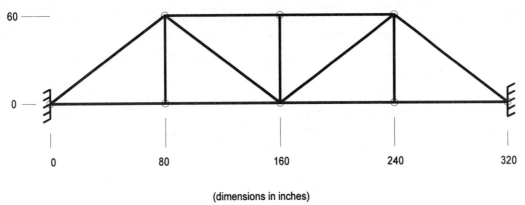

(dimensions in inches)

Figure 28 Frame geometry

Model Geometry

The easiest way to create this model is using a
sketched datum curve that contains all the
geometry of Figure 28. Start a new Pro/E part
[frame2D] using the default template. Change
your units to inch-pound-sec (IPS). Create a
sketched datum curve on the FRONT datum (see
Figure 29). Using the sketching constraints (not
shown in Figure 29), you should only need two
dimensions for this. Note that we don't create any
datum points in Pro/E - we will need a couple for
the constraints later and will create those
simulation features in MECHANICA.

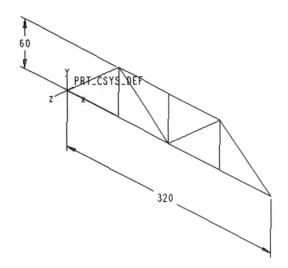

Figure 29 Sketched datum curve

With the geometry defined, transfer into
MECHANICA with

Applications > Mechanica

Accept the default model type.

Beam Elements

Create the beam elements directly from the
datum curve. Select (or use the *New Beam*
toolbar button)

Insert > Beam

Leave the name as **beam1**. Under References,
select **Edge/Curve** in the pull-down list.

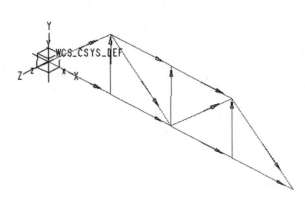

Figure 30 Selecting beam elements (arrow
shows BACS X-axis direction)

Separately select each of the datum curve segments. The frame highlights in red and a magenta direction arrow appears on each member. This is the X-axis direction of the BACS. If you want to (it is not necessary here), you can reverse this direction by left clicking on the member. With all selected, middle click.

Back in the **Beam Definition** window, beside **Material** select *More*. Transfer the material **STEEL** into the model and accept the dialog.

For the **Y Direction** select **Vector in WCS** and change the direction to **(0, 0, 1)**. This puts the BACS Y-axis for all beams in the WCS Z direction.

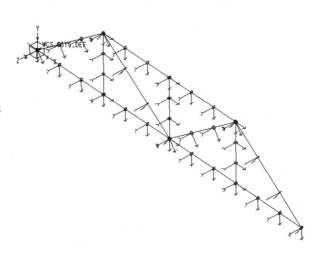

Beside **Section**, select *More > New*. Enter a name *[hcirc]* and a description ("hollow circle"). Beside **Type**, select **Hollow Circle** from the pull-down list. Enter an outer radius (R) of **1.5** and an inner radius (Ri) of **1.25**. Select *OK* here, and again in the beam definition window.

Figure 31 Beam elements created.

All beam elements are now defined as shown in Figure 31. For the following you might like turn off the display of the beam section, using *Setup Simulation Display > Modeling Entities* and deselecting the box beside **Beam Sections**.

Completing the Model

Constraints

We will constrain the two end points of the model. Select

New Displacement Constraint

Name the first constraint *[fixed_left]*. Set the **References** type to *Points*. Pick on the end of the model at the origin (left end). Observe the preselection filter and pop-up. Middle click to accept the end point. Back in the constraint dialog window, leave all constraints as **FIXED** and accept the dialog.

Create another point constraint at the other end of the frame in the same way. Name the constraint *[fixed_right]*. Leave all these constraints **FIXED**.

Loads

We will create two load sets. The first contains a uniform load applied to a curve. The second will be a load due to gravity. Keeping these in separate load sets means we can examine their effects separately.

Start with the uniform load. In the pull-down menu, select (or use the *New Force/Moment Load* button)

> *Insert > Force/Moment Load*

Set up a new load set called *[loads]*. Name the first load in the set **[download]**. Set the **References** type to *Edges/Curves* and pick on the third element from the left across the bottom of the frame (you may have to use the preselection filter here). Middle click. Set a **Total, Uniform** load with a component **-1000** in the Y direction (note this is WCS). *Preview* the load and select *OK*.

Now to apply the other load. In the pull-down menu, select (or use the *New Gravity Load* button)

> *Insert > Gravity Load*

A new dialog window opens where we define the gravity load. Name it **gravity** and create a new load set **gravload**. The acceleration is **-386.4** in the Y direction (remember we are using inches). See Figure 32. Select *OK*. A green icon appears at the origin (this may be obscured by the WCS and constraint symbol there)[3]. See Figure 33.

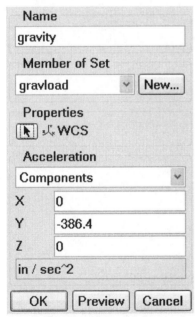

Figure 32 Defining a gravity load (IPS units)

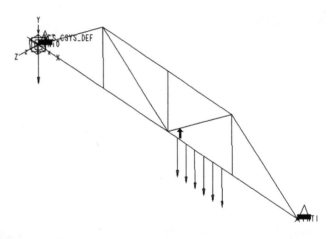

Figure 33 Model completed (beam section display turned off)

---

[3] Of course gravity acts on every element, not just the corner!

Analysis and Results

Set up a **Quick Check** analysis called *[frame2D]*. The constraint and both load sets (**loads** and **gravload**) should be highlighted. **Run** the analysis (don't forget to check the **Settings**). Check the **Study Status** for errors and warnings. If all goes well, change the analysis to a **Multi-Pass Adaptive**, 1% convergence. Set a Plotting Grid of *10*, and rerun the analysis. Open the **Study Status** window. The run converges in 3 passes. You might note the data for the resultant loads on the model. The resultant load for **loads** is -1000 in the Y direction. The resultant gravity load is -648, also in the Y direction, which is the weight of the frame. Note the maximum stresses and deflections for the two load sets. There is no torsion on any elements. The stresses due to the applied loads are about ten times greater than those due to gravity. The displacement in the Y direction is -0.043 inch for the applied load and -0.00488 inch for the gravity load.

Create three result windows showing displacement animations for the separate loads, and for a combined load. Set the deformation scale to 1500 in each window. These are shown in Figures 34 through 36 below. Note the continuity of slope of each beam through each connection. Each beam shows some bending.

Create similar result windows to show fringe plots of the **Total Von Mises** stress for each load set separately and for the combined loads. When you show all these simultaneously, you may want to set up the same legend scale in the three result windows. These figures are not reproduced here.

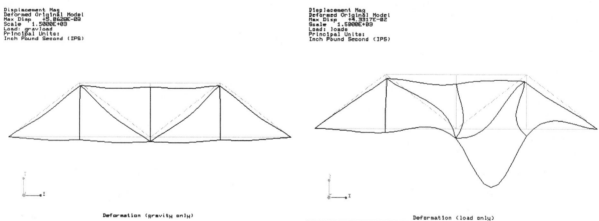

Figure 34 Deformation (gravity only)
Scale = 1500

Figure 35 Deformation (load only)
Scale = 1500

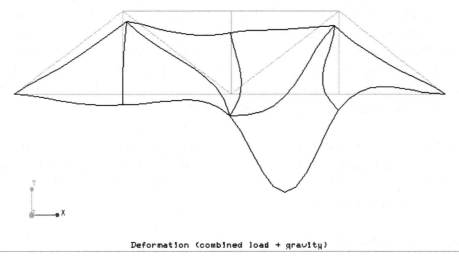

```
Displacement Mag
Deformed Original Model
Scale  1.5000E+03
Load: Combination
Principal Units:
Inch Pound Second (IPS)
```

Deformation (combined load + gravity)

Figure 36 Deformation (combined load + gravity) Scale = 1500

Model B - 3D Frame

We'll take the existing 2D frame and copy it to form another side of a 3D frame, then create some beam elements to connect the two frames. This is all done in Pro/E. In MECHANICA, we are eventually going to do something a bit different here: apply a displacement constraint to model a slumping foundation below one of the frame supports.

Modifying the Model

Bring up the frame part in Pro/E. To keep our models separate, do a *Save A Copy* and save the part using a new name, like *[frame3d]*. Erase the original part file and load the new one. To create the 3D frame, we will add to the original one.

First, we

> *Edit > Feature Operations > Copy > Move | Select | Dependent | Done*

and pick on the datum feature. Middle click and select *Done*. Then select

> *Translate > Csys*

Pick on the default coordinate system and pick *Z-axis*. Confirm the direction of translation and enter the offset distance *80*. Then select

Done Move

Accept all the remaining menus and accept the feature.

Now create some datum curves that connect the two frames. These will be in two new features. Create the first set of cross members (the lower ones) as a Sketched datum. Use the TOP datum plane for the sketching plane. In Sketcher, add additional references (for Intent Manager) at the vertices of the datums along the lower edge of both of the 2D frames. Then sketch five line segments that span the gap between the frames. No dimensions should be required for this sketch, since it is referenced completely to existing geometry.

Repeat for the upper cross members. This time, the sketching plane is a Datum-on-the-fly which is *Through* the top datum curves of each frame and parallel to TOP. Specify additional references for Intent Manager at the top vertices on both frames. Create the three curves in the feature to connect the top edges of the two frames.

The completed model at this point is shown in Figure 37. Note that this is in perspective view, obtained using

> *View > Model Setup*
> *Perspective*

and using the defaults.

We are ready for MECHANICA. Transfer there with

> *Applications > Mechanica*

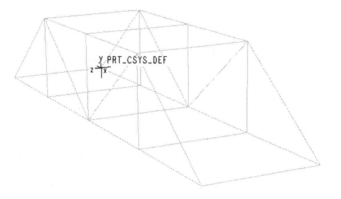

Figure 37 Perspective view of model datum curves

This brings the model into MECHANICA with all the previous elements, loads, and constraints already defined.

Creating Beam Elements

We need to create elements on all the new datum curves. For the new side frame, this can be done by editing the previous element definition. Pick on any of the original beam elements and in the RMB, select *Edit Definition*. The **Beam Definition** window opens. Select the button under **References**, and pick all the datum curves on the copied part of the frame (not the cross pieces). It is not critical that the direction arrows on the elements match the original. When these are selected (there should be 26 in all), middle click and then select OK in the **Beam Definition** window.

For the cross pieces, we will use the same section, but the orientation is different. Select the *New Beam* button in the right toolbar. Call the beam **beam2**. The Reference is *Edge/Curve*. Pick all eight cross members (it is easier to see these members when you are looking almost straight down on the model). These should all have the magenta arrows pointing parallel to the WCS Z direction. When selected, middle click. We will keep the same material (**STEEL**). The **Y Direction** for these members is defined by the **Vector in WCS (0, 1, 0)**, that is, the default. Keep the same beam section **hcirc** and accept the definition window.

The screen now displays all the beam elements. This display is getting a bit busy, so select

> *Setup Simulation Display > Modeling Entities*

and deselect the option **Beam Sections**. Under the **Settings** tab, you can also turn off *Load Value*.

Completing the Model

With this model, our primary aim is to examine the effect of a settled foundation under one corner of the frame. We need to set up some additional constraints, and modify the loading a bit.

To remove the load applied in the previous model, select in the display window, then in the RMB pop-up select *Delete*.

In the model tree, highlight the loadset named **loads**, and select *Delete*. Confirm the deletion. All we should have in the model is **gravload**. The icon for this should still be shown on the WCS origin.

Constraints

We need to constrain the new front corners of the frame. Select

> *New Displacement Constraint*

Name the first constraint *[front_left]*. Make sure this is in ConstraintSet1. Set **References** to *Points* and pick at the left end of the frame on the new front members. Leave this point totally **FIXED**. Repeat this procedure for the other corner, naming it *[front_right]*.

The model is now complete. See Figure 38 below (note that this is a perspective view, which makes interpretation of wireframe models a lot easier). You can just make out the gravity arrow at the far back corner.

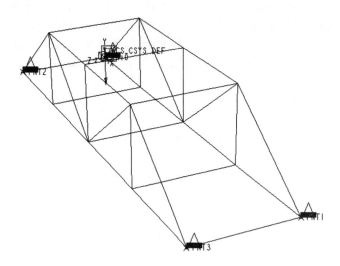

Figure 38 Perspective view of complete 3D frame (Model B) with gravity load

Analysis and Results

Set up a **Quick Check** analysis called *[frame3d]*. Check your *Settings* and *Start* the analysis. Scan through the *Study Status* window. There are 34 beam elements. Note the total resulting load on the model is -1687 lb in the Y direction - the total weight of the frame. Assuming there are no errors found, *Edit* the analysis to produce a **Multi-Pass Adaptive** analysis (1% convergence). Leave the maximum edge order at 6. Set the **Plotting Grid** to 10. Run the new analysis, deleting the existing output files, and open the *Study Status* window. Convergence is on the 3rd pass with a maximum edge order of 6. Note the maximum Y displacement is -0.0063 in.

Create a result window showing a deformation animation. Be sure to select the design study **frame3d**. Set a **Deformed Scale** of 2500. The deformed shape is shown below in Figure 39. The animation of this looks like the frame is melting!

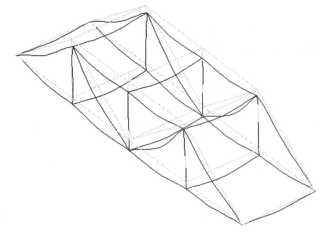

Figure 39 Deformation of Model B under gravity load

Displacement Constraint

Now we will modify the constraint at one corner to simulate a settling foundation for the frame. This represents a forced displacement in the Y direction for the constraint.

Click on the front, right corner point constraint (or select the constraint **front_right** in the model tree). In the RMB pop-up select ***Edit Definition***. Change the Y translation constraint to the third button - **Prescribed**. Enter a value of ***-0.1*** (not very much compared to the height of the frame!) Accept the dialog.

The constraint icon does not indicate this change in the constraint. Is it in the model? Go to the model tree, expand the Loads/Constraints and ConstraintSet1 entries, right click on **front_right** and select ***Info***. This information window shows the nature of the constraint.

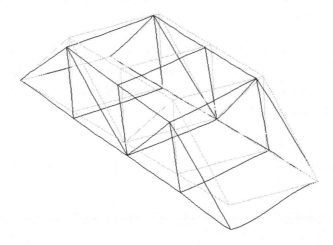

Figure 40 Deformation of Model B with displaced support on front corner

Start up the previously defined static analysis. Delete the output files and open the ***Study Status***. The maximum Y displacement is at the constraint, -0.10 inch (no surprise there!). Create a new deformation animation result window. Use a scale factor of 200. The frame looks like Figure 40. Note the curvature in the crossbeam at the right end of the frame and recall that we have set both rotation constraints to **FIXED** on all of these point constraints. You might think about this for a minute, and then find out what happens if we free these rotation constraints and run the analysis again.

Finally, create a result window that shows the total Von Mises stress fringe plot. In this result window, you will note that the bending stress is very low in all elements except the crossbeam entering the displaced point. What happens to the stress if you edit all the constraints to free rotations around all axes? What do you think about the way we have applied the constraints?

Summary

This has been a busy lesson and we have covered a lot of material. Beam models can be the simplest in terms of geometry (using just datum points and curves), but possibly the most difficult to set up in terms of modeling parameters required. The most difficult of these parameters, particularly in 3D, are related to the problem of determining beam orientation. In addition, we have not dealt with asymmetric beams, like channels or angles, or curved beams. Before you try that, you should consult the MECHANICA documentation and study the sections on the BSCS (Beam Shape Coordinate System) and the BCPCS (Beam Centroidal Principle Coordinate System). The idea of beam releases is also probably a new concept, and their use in modeling will require some additional study.

Beam elements do not need to be used in isolation. They can also be used in conjunction with solid and shell elements (in 3D). You must be careful, however, about joining the end of a beam to a solid. Recall that a beam end point has six degrees of freedom, while a solid node has only three. Therefore, transfer of a beam bending moment from a beam into the solid is a complex and advanced process.

To gain more practice with beams, you are encouraged to try some of the exercises below.

Questions for Review

1. What does BACS stand for?
2. How is the X-axis of the BACS determined?
3. How are the Y- and Z-axes of the BACS determined? That is, what is the relation of the BACS to the WCS?
4. When a load is applied to a beam, in what coordinate system are its components specified?
5. What is the BSCS? What parameters are used to define it? Relative to what?
6. What standard beam sections are available in MECHANICA? How can you determine their section properties (like I_{ZZ}, I_{YY}, and so on)?
7. How can you determine the direction of an element's X and Y directions?
8. In our diving board problem, how would you model the case of a person standing on one of the corners at the tip of the board (causing it to twist)? Discuss both loads and constraints for this scenario.
9. Find out if and where MECHANICA writes any data to a file associated with the shear and moment diagrams.
10. What is the MECHANICA general guideline for the beam element geometry?
11. Is it possible to create a single distributed load that spans several elements?
12. Is it possible to have two or more distributed loads acting on the same element, say in different planes (e.g. a linear distributed load in the XY plane, and a quadratic load in the XZ plane)?
13. Is it possible to have both point loads and distributed loads acting on the same element?
14. Can a point load act in the center of a beam element?
15. Is it possible to model a tapered beam in integrated mode? If so, how do you set it up?
16. What sign convention does MECHANICA use to draw shear and bending moment diagrams? Is this the same one you usually use?
17. What is the purpose of a beam release, and how is it applied? How must the beam element be created to apply a release?
18. Do you have to release *both* beam elements meeting at a point?
19. Assume there are two collinear beam elements that meet at a point. Describe the physical situation that would result in the following beam releases for one of the elements at that point: (✔ = released)

	Translation Released			Rotation Released		
	X	Y	Z	X	Y	Z
Case 1		✔				
Case 2					✔	✔
Case 3				✔		
Case 4				✔	✔	✔
Case 5		✔				✔

20. What is the easiest way to create a frame?
21. For our 3D frame, what is the minimum set of constraints required for static analysis?
22. What happens if you try to show a shear force and bending moment diagram of the 4 beam elements across the bottom edge of our 2D frame model?
23. What happens if you try to show a shear force and bending moment diagram of beams whose X-axes are not co-linear (for example, two elements that form an "L")?

Exercises

1. For the beam and loading shown below, find the maximum bending stress. Plot the shear and bending moment diagrams. Redo the problem assuming pinned ends. The cross section is a hollow rectangle, 4" high and 3" wide with a wall thickness of 0.25". The material is steel. The load distribution on the right is quadratic.

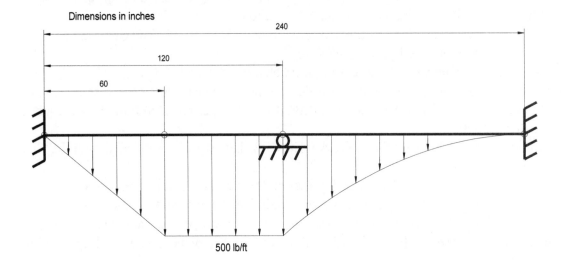

2. A transmission tower is shown below (2D model only). During a hurricane, the loads on the tower caused by the cables are deflected 30° from the vertical. Using a frame model, find the maximum stress and deflection in the tower. The steel cross section is a hollow circle, OD 8.0 cm, with a wall thickness of 5.0 mm. Are any members in danger of buckling?

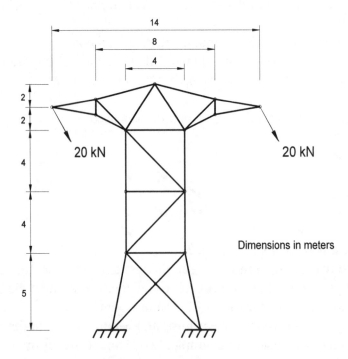

Synopsis

Constraints using cyclic symmetry; spring and mass element idealizations; creating measures; modal analysis (natural frequency and mode shapes); contact analysis in a simple assembly

Overview of this Lesson

In this lesson, we will examine a variety of modeling tools to round out our exploration of MECHANICA Structure. As mentioned earlier, cyclic symmetry is a close relative of axisymmetry and we will look at how it is set up for a solid model. In another model, we will set up an example of the use of idealizations of linear springs and a point mass (with a gravity load). In this example, we will see how to create our own measures to extract precise information from the results. The same model will be used to introduce procedures for carrying out a modal analysis. This analysis will determine natural frequencies of vibration and the associated mode shapes. Finally, we will look at how a contact analysis can be used in a simple assembly.

As usual, there are some Questions for Review and a few Exercises at the end of the lesson.

Cyclic Symmetry

An enhancement introduced in MECHANICA a few releases ago was a new kind of constraint that can be used for models that have cyclic symmetry. This is similar to axisymmetry in which the model is obtained by revolving a planar section (or curves defined on a plane) around an axis of revolution. Axisymmetric models are determined by 2D geometry and were covered in Lesson 6. In a model with cyclic symmetry, a 3D geometric shape is repeated identically an integer number of times around the axis in an equally spaced pattern. The geometry is not continuous (and therefore not axisymmetric), but cyclic. If the loading and constraints are also cyclic along with the geometry, then it makes sense that we should be able to analyze a single portion of the overall geometry. This is illustrated in Figures 1 and 2. The first figure shows a complete centrifugal fan impeller (somewhat simplified!). Because of symmetry about a horizontal plane, we can cut the fan in half, as seen in the second figure. If the loading is identical on all the blades (perhaps due to the centrifugal load), we should be able to analyze a single blade by properly isolating it and applying constraints that capture the repetitive or cyclic symmetry. We will return to this model a bit later.

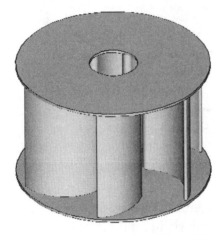

Figure 1 A centrifugal fan impeller

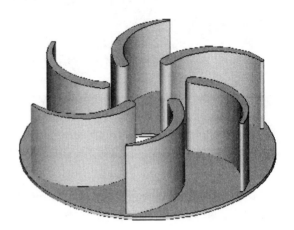

Figure 2 Lower half of fan impeller

The major requirement for using cyclic symmetry is that the following items are all cyclic:
- geometry
- applied loads
- constraints
- material type and orientation

MECHANICA will try to determine the axis for the cyclic symmetry automatically (as in the example following). If it cannot do this, you will be prompted to identify the axis yourself.

To demonstrate the procedure, we will revisit a model we dealt with earlier. This is the pressurized axisymmetric tank we treated using a 2D axisymmetric solid model. Note that the tank is fully axisymmetric, which is not required for cyclic symmetry. However this is a simple model to create and gives us a chance to compare results with another analysis method. The completed cyclic symmetry model is shown in Figure 3. This is a 30° wedge-shaped portion of the tank.

Model Geometry

In Pro/E, bring in the solid model **axitank** that we created in Lesson #6. Use *Save A Copy* to create a new copy of the part called **cycsym**. Erase the original part and bring **cycsym** into the Pro/E session. The model has a 90° revolved protrusion representing one-eighth of the tank. Use *Edit* to modify the feature so that its angle is just 30°. Recall that the units for this model are IPS.

Transfer into MECHANICA with

Applications > Mechanica

All the previous modeling entities (loads, constraints, etc) are still there[1]. Before proceeding with the cyclic model, delete the existing constraints on the vertical surfaces. This is easiest to do using the model tree. Open this up and expand the entries listed under **Loads/Constraints**. We want to delete the constraints **XYface** and **YZface**. Select these one at a time and, holding down the right mouse button, select **Delete** from the pop-up window. The constraint **XZface** on the lower horizontal surface is still required.

Cyclic Constraints

To define the new constraints it will be necessary to have a cylindrical coordinate system at the origin of the datum planes. Create that now using the ***Datum Coordinate System*** tool, selecting ***Type(Cylindrical)***.

Use the procedures discussed earlier to create a cylindrical coordinate system as shown in Figure 3. The lower front edge of the model corresponds to the **Theta = 0** direction, and the axis of the revolved protrusion to the **Z-axis**. We must declare this as the current system using (in the pull-down menu)

Edit > Current Coordinate System

Pick on the created cylindrical system (probably **CS0**). It will turn green.

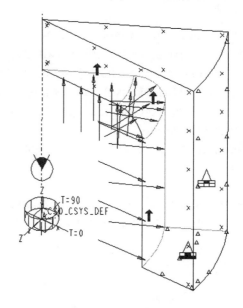

The constraints on the two vertical faces of the model are cyclic. As our pie-shaped model is repeated (12 times) to form the upper half of the tank, the solution values on the two vertical faces must match up. Set up the cyclic constraint using (starting in the pull-down menu) (or use the new ***New Symmetry Constraint*** toolbar button):

Figure 3 Completed cyclic symmetry model

Insert > Symmetry Constraint

Enter a constraint name *[cyclic]* (in ConstraintSet1). Set the **Type** to *Cyclic*. Under **References**, select the button below **First Side**, and pick on the front vertical face, then middle click. Now select the button below **Second Side** and pick on the back vertical face; middle click again. Notice that at the bottom of this dialog window, the **Axis** area is grayed out - MECHANICA was

---

[1] If you have been carefully following the lessons, the model should be set up as the solid model used in Lesson #6 with constraints for mirror symmetry.

able to determine automatically where the axis is located (intersection of the two planar surfaces). Accept the dialog. Constraint symbols (small x's) will appear on the two surfaces and the cyclic symmetry icon appears on the axis of the model as in Figure 3.

This model is not yet fully constrained against free body motion. What motion is still possible and consistent with the existing constraints? This is why we needed the new cylindrical coordinate system. Select

New Displacement Constraint

Name the constraint *[outer_face]* (in ConstraintSet1). Select the button under **Surfaces** and pick on the outer curved face[2]. The constraints to set here are: **FREE** the translation in R and Z, and **FIXED** for translation in Theta. Recall that rotational constraints are ignored for solid models.

Finally verify or use *Materials* to specify the material **STEEL** for the model. The model should now appear as shown in Figure 3.

Analysis and Results

Go to the *Analyses/Studies* window (you can delete the study **axitank_solid** if desired). Set up and run a **QuickCheck** analysis called *[cyclic]*. Use ConstraintSet1 and LoadSet1 (that contains the 1000 psi pressure load). AutoGEM will create 12 solid elements. Assuming all goes well, change to a **Multi-Pass Adaptive** analysis with **10%** convergence and a maximum polynomial order 6. When you run this, open the *Study Status* window. The analysis should almost converge on pass 6. Note the maximum Von Mises stress (about 4930 psi), and the maximum deflection $\Delta y_{max} = 0.00026$ in. Compare these to the results obtained in Lesson 6 - they are very close.

Create a couple of result windows showing the convergence of the Von Mises stress and the strain energy. These are shown in Figure 4 below. Pretty good convergence here.

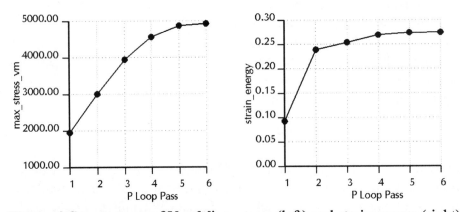

Figure 4 Convergence of Von Mises stress (left) and strain energy (right) in tank with cyclic symmetry

---

[2] You can also pick the top surface, and the inside surfaces. Come back later to add these and see if there is any effect on the model or computation.

Create result windows showing a Von Mises stress fringe plot and a displacement animation. These are shown in Figures 5 and 6. Note the stress pattern on the side (cyclic) faces are identical and very similar to the axisymmetric model in Lesson 6, and is uniform around the axis of rotation along the round. In the deformation animation, the deformation is also as expected and consistent with what we saw before.

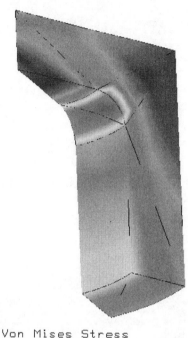

Von Mises Stress

Figure 5 Von Mises stress in cyclic model

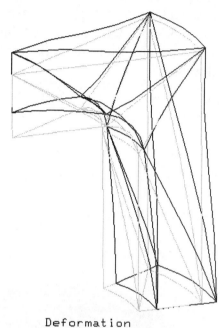

Deformation

Figure 6 Deformation of cyclic model

Let's return to the example mentioned at the beginning of this section - the centrifugal fan problem. In Pro/E, we can isolate one of the blades using judicious cuts. Beware that the surfaces where cyclic constraints are to be applied must be identical shapes - the cyclic "instances" of the blade and end plate combination must fit perfectly around the whole fan. The included angle between the cuts must be obtained as 360° divided by a whole number, in this case, 6. This is easy to set up in Pro/E if you use a datum curve to define the desired shape of the cut on the end plate and then create a rotated copy (around the central axis) of the datum curve. You then create the cut using the *Use Edge* option in Sketcher and picking the datum curves.

The resulting geometry and model are shown in Figure 7. The cyclic symmetry constraints are placed on the S-shaped cut faces of the end plate. Could a shell be used for this part of the model? Depending on the geometry of the end plate cut, you may have to identify the cyclic symmetry axis for this geometry. The next constraint is due to symmetry on the top surface of the blade in the model to prevent rigid body translation along the axis. Finally, the model must be constrained against rigid body rotation around the central axis. This is easily done with a cylindrical coordinate system in the same way we handled the previous tank problem. This should probably be applied on the inner surface of the end plate. The model is loaded with a centrifugal load (notice the symbol in the upper left corner of the figure) that appears on the axis of rotation of the fan.

Some results of running this model (326 solid elements) are shown in the figures below. Note the maximum stress levels are near the boundaries and at the junction of the blade trailing edge and the end plate, which is a reentrant corner. The maximum Von Mises stress convergence graph (not shown here) also displays evidence of a problem. We might have expected this! This model would need some re-work before providing good results.

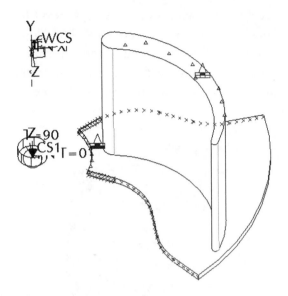

Figure 7 Model of fan blade using cyclic symmetry

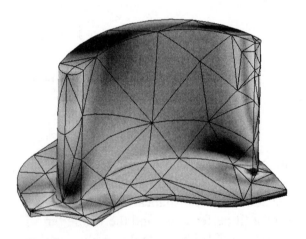

Figure 8 Von Mises stress on fan blade in model with cyclic symmetry. Note hot spot at blade root near trailing edge.

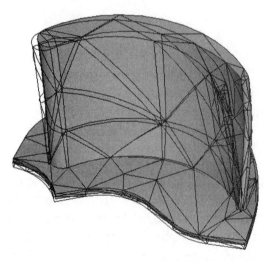

Figure 9 Deformation of fan blade and end plate

Springs and Masses

In this section we will look at two more idealizations - springs and masses. Springs can be either extension/compression springs or torsion springs that connect two points, or one point and the fixed ground. Masses can be either point masses or can be given inertial properties that will affect their rotation.

We will examine a simple model composed of two short beams cantilevered out from a wall. The 6" long aluminum beams have a solid circular cross section (diameter 0.5"). The free ends of the beams are connected by a linear spring. Another spring supports a mass (we will model it as a point mass). The only load on the system is due to gravity. The physical system is depicted in Figure 10.

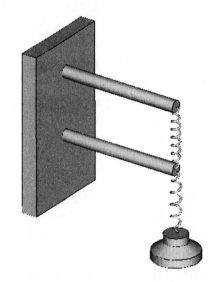

All springs in MECHANICA are linearly elastic. There are two kinds of springs: **Two Point** (connecting any two points in the model), and **To Ground** (connecting any single point directly to the fixed ground). *Two Point* springs can come in any orientation (specified somewhat like a beam), whereas *To Ground* springs are always oriented parallel to the WCS. The spring properties required are its *Stiffness* (extensional, torsional, or mixed) and its *Orientation* (for Two Point springs only). It is possible to constrain a model entirely using To Ground springs, rather than having "hard" constraints. You must be careful in that case that every relevant degree of freedom has some spring stiffness associated with it.

Figure 10 Model with spring and mass entities

Masses also have a number of options. The mass element is applied at a point (which can be created on the fly). We are not restricted to point masses, though, since the mass element can be given inertial properties that would respond to rotation of the model (for example in a centrifugal loading case).

Model Geometry

Our idealized model of the beam/spring system will consist of two beam elements defined on datum curves. Start Pro/E and create a new part called *[twobeams]*. Set your units to IPS. On the **FRONT** datum plane, create 5 datum points with the dimensions shown in the figure at the right. These are numbered **PNT0** through **PNT4**. You might like to keep **PNT4** in a separate feature since we will be deleting just this point a bit later in the lesson.

Create a couple of datum curves to connect the top pairs of datum points (use *Thru Points > Single Point*). The completed Pro/E model is shown in Figure 11.

Bring the model into MECHANICA with

Applications > Mechanica

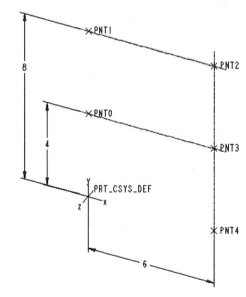

Figure 11 Pro/E model of system

Accept the default model type (3D).

Creating the Elements

This model will have three kinds of idealized elements. First, we'll create the beam elements. Select (or use the *New Beam* button)

Insert > Beam

Use the default name **Beam1**. In the **References** area, select **Edge/Curve** in the pull-down list. Pick the two datum curves. On each beam element, the purple arrow indicates the beam's X-axis points parallel to the WCS X-axis. In the **Materials** area, select *More* and add **AL2014** to the model and select *OK.* For the **Y Direction**, use **Vector in WCS** and keep the default direction (0, 1, 0). In the **Section** area, select *More > New*. Name the section [circle]. Select *Type(Solid Circle)* and enter a radius of **0.25**. Accept all the dialogs. The beam section icons appear on the model in light blue.

Now create the springs. Select (or use the *New Spring* button)

Insert > Spring

Call the first spring **spring1** (default). The default type is a *Simple* (ie no lateral stiffness), point-to-point spring. Select the button under References and pick on the points at the ends of the two beams (PNT2 and PNT3 in Figure 11). Enter an extensional stiffness of **2000** (note units of lbf/in). See Figure 12. You can also obtain this stiffness from a Pro/E model parameter (see Exercise #7). Accept the dialog. A spring icon appears on the model.

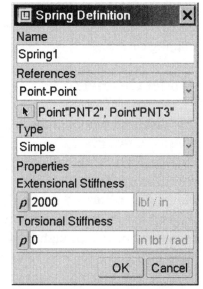

Figure 12 Defining a simple spring

If you pick an *Advanced* spring type, things get quite a bit more complicated. You can set lateral stiffness as well, relative to a coordinate system fixed to the spring. In this case, you must also worry about the orientation of the spring. This is given in much the same way as the orientation of a beam element: the spring X-axis connects the two points and you specify the direction of the spring Y-axis in the WCS.

The last **Type** option is **To Ground**. This can be used to provide "soft" constraints to points on a model that offer resistance in proportion to deflection instead of "hard" constraints that restrict motion totally.

Create the second simple spring between points PNT3 and PNT4. This is also a simple, point-to-point spring. The extensional stiffness is **1000.** Accept the dialog.

Now to create the mass. In the pull-down menu, select (or use the *New Mass* toolbar button)

Insert > Mass

Note the defaults. Call it **mass1** and pick on the lower point (**PNT4**). Middle click. We want a weight of 10 lb. The mass is (weight/gravity = 10 / 386.4 =) **0.026**, since in the IPS system g = 386.4 in/sec^2. What are the units of mass in IPS?

Note that by picking an advanced mass type, we can also specify moments of inertia, which might be important if the mass was going to rotate. Accept the dialog and a dark blue mass icon appears on the model.

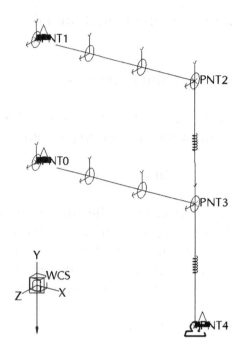

Figure 13 Completed FEA model

Loads and Constraints

We'll cantilever the beams out from the wall:

Constraints > New > Point

The constraint name is *[fixed]* (in ConstraintSet1). Set **References** to **Points** and pick the two points of the beams at the wall (**PNT0** and **PNT1**). Leave all degrees of freedom fixed and accept the constraint.

The gravity load is applied with (or use the **New Gravity Load** toolbar button):

Insert > Gravity Load

Note that gravity is relative to the current coordinate system (WCS). Enter a name *[gravity]* in LoadSet1 and enter a value of **-386.4** (in/sec^2) in the Y direction.

Open the **Analyses/Studies** window. If you select the **Info > Check Model** command, no errors will be found. However, if you try to run the model now, it will fail with an "insufficiently constrained" error message. Why? Since the mass is just hanging on the end of the spring, in addition to the desired vertical motion it also has two rigid body degrees of freedom (translation in horizontal X and Z directions). For our purposes, we can constrain it to move only in the Y direction. Do that now (make sure the constraint is in ConstraintSet1)[3]. Apply the constraint to

---

[3] Come back later and see if converting the lower spring to an Advanced type and adding Kyy and Kzz properties solves this problem instead of adding the extra constraint.

PNT4 and leave all the degrees of freedom as **FIXED** except for translation in the Y direction. The completed model should look like Figure 13.

Analysis and Results

Set up a **Quick Check** static analysis. *Run* the analysis (you can ignore the warning about stress concentrations) and open the *Study Status* window. Note the Model Summary at the top (2 springs, 1 mass, 2 beams). Assuming there are no errors, *Edit* the analysis to produce a **Multi-Pass Adaptive** analysis with 5% convergence, maximum polynomial order 6. Run the MPA analysis. Open the *Study Status* window. The run converges on pass 2 with max edge order 4. The resultant load in the global Y direction is -10.28 lb. The extra 0.28 lb is due to the weight of the beams! Check out the stresses and displacements. What is the maximum displacement, and where does it occur? It is probably the deflection of the point mass. Create a result window to show a deformation animation. The springs and the mass are not displayed. Change the window definition to:

> *Display Type(Fringe)*
> *Quantity(Displacement)*
> *Component(Y)*

and show the window. The tip displacement of the lower beam is greater (in magnitude) than that of the upper beam. What are the deflections of the two beam tips exactly? And what about the mass? Here's how we can find that out by defining our own *measures*.

Defining Measures

In the pull-down menu, select (or use the *Define Simulation Measure* toolbar button)

Insert > Simulation Measure

Select *New* and call the first measure *[defy_top]*. Click the *Details* button and enter a description. Select the following in the pull-down lists:

> *Quantity (Displacement)*
> *Component (Y)*
> *Spatial Eval (At Point)*

See Figure 14. Select the Points button and click on the end point on the top beam. A small icon appears at the tip of the beam. *Accept* the dialog.

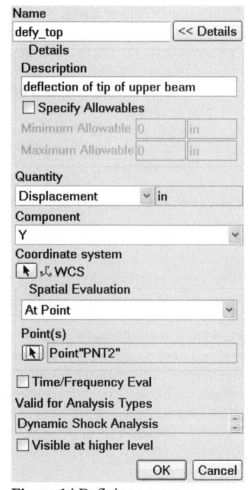

Figure 14 Defining a measure

Copy this definition to create a new one; open it with *Edit*. Change to a new name *[defy_mid]*. Keep the same settings but select a new point on the end of the lower beam (the middle of the three points on the right). Accept this definition.

Copy **defy_mid** to a new measure *[defy_mass]*. *Edit* this definition and select the mass point.

We now have three of our own measures. Unfortunately, custom measures are not retroactive - they are only computed during the analysis run and not after (plan ahead!). Go to the *Analyses/Studies* window and *Start* the analysis again. Open the *Study Status* window. At the bottom of the Measures list in the summary we find our three measures. They are **defy_top** = -0.0101 in, **defy_mid** = -0.0124 in, and **defy_mass** = -0.0224 in. What is the extension of the lower spring? What is the force in the lower spring? Does this value make sense? How much of the 10 lb weight is carried by the upper beam?

This model is the basis for an exercise at the end of the lesson, so make sure you save it.

Return to Pro/E for the next section of the lesson.

Modal Analysis

As you probably know, when a structure is excited by a dynamic load at close to its natural frequency, you can expect trouble. Therefore, it is important to be able to predict what the natural frequency is. Furthermore, for all continuous systems, there are a number a frequencies of vibration that can occur naturally. Each frequency has associated with it a characteristic deformed shape, called the mode shape. The modes are numbered, with mode 1 having the lowest frequency (it is called the fundamental mode). We are usually concerned with the modes that have the lowest frequencies. In MECHANICA, we can find these frequencies and mode shapes using *modal analysis*. We will investigate this form of analysis using the beam/spring model from the previous section.

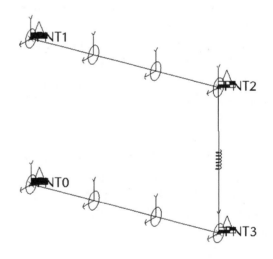

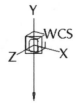

Figure 15 Model for modal analysis of beam/spring system

Using *Save A Copy*, make a copy of the **twobeams** model created in the previous section. Call the new model *[twobeams_modal]*. Erase **twobeams** from the session and load the new model. Delete the datum point where we put the mass (**PNT4** in Figure 11) - this also will remove associated modeling entities. Bring the modified model into MECHANICA.

Setting up the Model

We have no other geometry to create. However, we are interested primarily in the motion in the XY plane, so let's constrain the tips of the beams for this. Select

New Displacement Constraint

Name the constraint *[beamends]* (in ConstraintSet1). Set **References** to *Points* and select the two points at the ends of the beams. **FREE** the translations in X and Y, and the rotation in Z. See Figure 15. (You should come back later and delete these end constraints to see what happens.)

Defining the Modal Analysis

Set up the analysis with

> *Analysis*
> *Mechanica Analyses/Studies*
> *File > New Modal*

Call the analysis *[twobeams_modal]*. Enter a description. There are quite a few differences in this window from what we have seen before. First, notice that there is no indication of the load set. For modal analysis, you don't need a load set. There are options for constrained and unconstrained analysis. Leave the default for this (consult the on-line documentation for further discussion). There are several tabs at the bottom. Under the **Modes** tab, you can select how many modes you want MECHANICA to find - the default

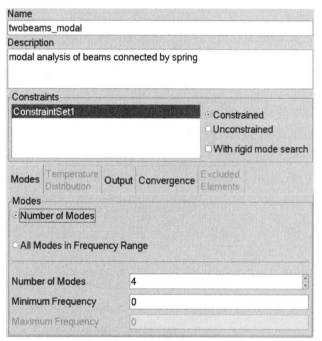

Figure 16 Defining a modal analysis

is 4. If you are concerned about a special frequency (for example, due to the speed of a rotating machine), you can search for modes in a near-by frequency range. Under the **Output** tab, change the **Plotting Grid** to 10. Under the **Convergence** tab, set a **Multi-Pass Adaptive** analysis with 5% convergence. Notice that this convergence is on frequency by default. See Figure 16 for the completed window. Accept the dialog window.

Now *Run* the analysis **twobeams_modal**. Open the *Study Status* window. The run converges on pass 3 (maximum edge order 6). The four natural frequencies are listed. They are

mode #1	389.7 Hz
mode #2	1068 Hz
mode #3	2387 Hz
mode #4	2408 Hz

We'll create four windows to show the four mode shapes. Select

Analysis > Results > Insert > Result Window

Call this first window *[m1]* and select the design study **twobeams_modal** in the output directory.
In the next window, select mode **1** with **Scaling** set to 1. Enter a title *[Mode #1]*. Set *Display
Type(Model)*, and *Quantity(Displacement)*. Under **Display Options**, set up an animation with a
scale of 10%. Check the options **Overlay Undeformed** and **Alternating**.

Copy the current window **m1** to a second window *[m2]*. Accept the window definition as is
except for selecting mode 2 and changing the title to *[Mode #2]*. Accept the edited definition.

Continue with this procedure to create windows for mode #3 and mode #4.

When all four windows are created, show them all at once. Each result window shows the mode
number and frequency. Be careful about your viewing direction for these windows - observe the
coordinate system triad in each window. The display should look something like Figure 16.

The first mode shows the beams moving up and down in unison (spring not stretching?). Mode 2
shows the beams moving in opposition (spring definitely stretching!). Mode 3 shows the beams
moving in unison with an S-shape. All of these are in the vertical XY plane. Mode 4 shows the
beams bending in opposition in a horizontal plane with the tips almost stationary.

You might explore the effect of different constraints applied to the beam end points. For
example, what happens if you **FREE** the rotations around X and Y? What happens if you
remove the constraints on the tips of the beams?

Close this model and return to Pro/E.

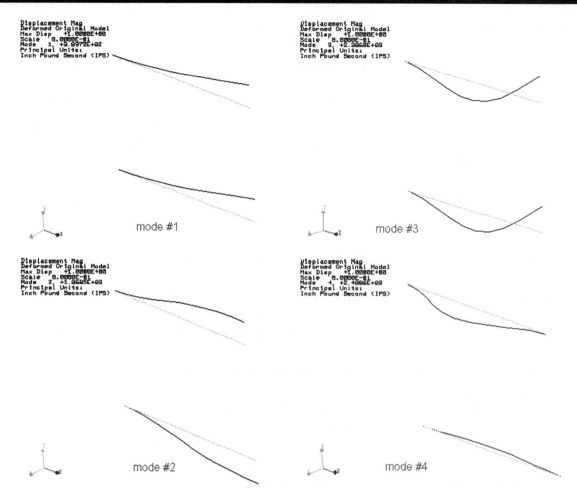

Figure 17 First four mode shapes of beam/spring system

Contact Analysis

Pro/Engineer is noted for its ability to create and manage assemblies. These assemblies are easily brought into MECHANICA for stress analysis. Be warned that (despite what we will see here) analysis of assemblies requires very advanced modeling techniques and understanding. One of the requirements for analysis of assemblies is the proper modeling of the contact between adjacent parts. This will be illustrated using the simple assembly shown in Figure 18. This consists of an aluminum base plate (approx. 24" X 10"), a steel pin (diameter 3"), and an aluminum connector (diameter 6"). The holes in the lugs on the base plate and on the connector are the same as the pin diameter. The dimensions of the lugs on the base plate and the cut out on the connector are not critical to what we are going to do - an approximate geometry here is satisfactory. When assembled, an upward force will be applied to the connector. To take advantage of symmetry in the geometry and loads, a quarter model can be created in Pro/E using a cut created in assembly mode. See Figure 19 (Note that this model is in the positive quadrant - view is from top - left - rear; observe the coordinate systems in Figure 19).

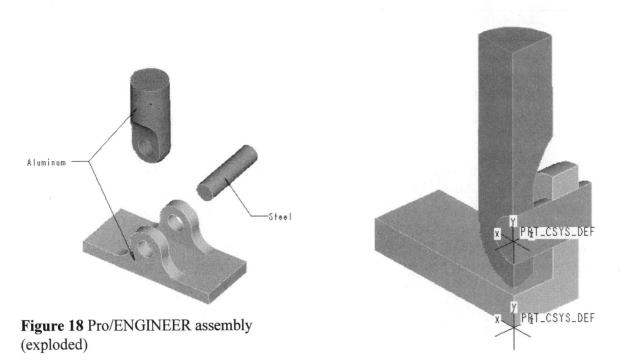

Figure 18 Pro/ENGINEER assembly
(exploded)

Figure 19 The FEA model

After you have created the parts, create an assembly in Pro/E. Make sure the units for each part
and the assembly are consistent (IPS). Use a cut in the assembly to extract the quarter model
(Figure 19) from the entire assembly. When the quarter model is ready, bring it in to
MECHANICA and assign materials to the parts. Select

Applications > Mechanica

Accept the default model type. You may notice some new toolbar icons have appeared since we
are looking at an assembly. Can you find them?

Using *Define Materials*, bring the materials **AL2014** and **STEEL** into the model. Then, with
AL2014 highlighted, select

Assign > Part

Click on the two aluminum parts (they highlight in red); middle click. Select **STEEL** and assign
this to the pin. Accept the dialog.

Constraints are applied to the two symmetry planes, and the lower surface. Starting with

New Symmetry Constraint

apply appropriate constraints (mirror symmetry) to the two cutting planes arising from symmetry.
Name these *[XYface]* and *[YZface]*. These involve fixing the translation perpendicular to the
surface, and freeing the other two translations. These are handled automatically by the mirror
symmetry constraint. Since the model will use solid elements, the rotational constraints don't

matter. Note that surfaces of all three parts should be included in each constraint. Finally, apply a translational constraint to the bottom of the base to prevent rigid body motion perpendicular to that surface.

Now apply an upward load:

New Force/Moment Load

Create a load on the top surface of the connector (**Uniform**, **Total Load**, **5000** lb upward). The model should appear as shown in Figure 20.

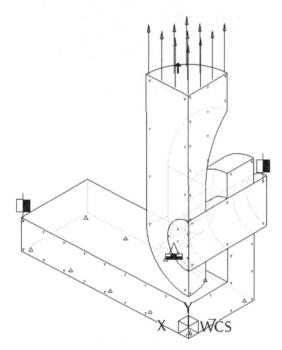

Figure 20 Model with constraints and loads complete

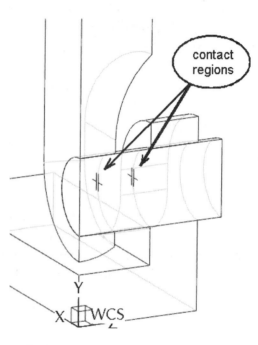

Figure 21 Contact regions defined between the pin and hole surfaces

Creating Contact Regions

If we perform an analysis of this model now, MECHANICA will "weld" the surface of the steel pin to the aluminum holes. This essentially creates a continuous solid whose material properties change as you cross the pin-hole surface. However, we know that because of the construction of the assembly and the applied load, as the load is applied, these surfaces might actually separate (create a gap). This is because contacting surfaces cannot support a normal (tensile) stress. We can define contact surfaces that MECHANICA can monitor for this - the surfaces are free to move apart (normal to the surface) to cause a gap to be created, but will not penetrate the other part. Contact will always give rise to compressive normal stresses. To create these contacts, in the pull-down menu select

Insert > Connection > Contact Region > Create > Face/Surface

and click on the cylindrical surface of the pin, then (using the right mouse button for pre-selection) click on the hole surface of the base part that mates with the pin. A small yellow contact region symbol will appear along with a couple of measure icons. Do the same where the pin passes through the hole on the connector. The two contact region symbols are shown in Figure 21. Use the *Info > Review Contacts* command to confirm the surfaces are correct. If you select one of the measure icons, you can use the RMB pop-up menu to select Edit Definition. This will let you get more information about that measure. There are actually four measures defined: for each contact region, measures were automatically created for contact area and maximum contact pressure. The icons may be overlapping on the screen.

Now set up and run a **QuickCheck** analysis called *[contact]* to see if we have any errors in the model. Notice the option (it should be checked) **Include Contact Regions**. At the bottom of the analysis definition window, we can set the number of **Load Intervals** to be used to apply the load. This is to account for the fact that the actual contact between surfaces in a complicated assembly may be made and broken as the load is applied. Furthermore, as contact is a non-linear problem, MECHANICA must do some iteration here to determine the size/shape of the actual contacting regions. For now, use a single interval. (Come back later and try five equally spaced intervals and compare results.) *Run* the analysis. AutoGEM creates 150 or so solid elements. Assuming no errors, change the analysis to a **Multi-Pass Adaptive** analysis (10% convergence, max order 6). The run will take a few minutes and should converge on pass 4 or 5. In the *Study Status* window, note that the contact area and maximum pressure is given for each of the contact regions.

Create result windows for the Von Mises stress and a deformation animation. Zoom in on the contact regions in the deformation window to observe the separation of the surfaces.

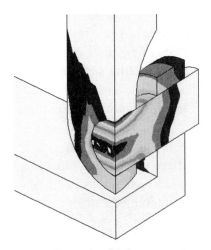

Figure 22 Von Mises stress fringe plot

Figure 23 Deformed shape showing gaps around pin

Create another result window showing the normal stress component in the direction of the load (the YY component). Check the option to show the model in the deformed shape. The result window is shown in Figure 24 (deformation scale is about 300). Note the gap that has been created around the pin, and the continuity of the normal stress where the surfaces remain in contact. You might like to examine the YY stress on an XY cutting plane through the model.

Finally, create result windows to show the convergence of the contact pressure (**contact_max_pres**) and total contact area (**contact_area**) for the two regions. These are pretty well behaved.

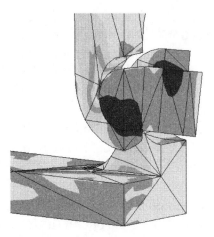

Figure 24 Normal stress (YY) fringe plot

Summary

This lesson has covered a grab-bag of miscellaneous modeling topics.

Cyclic symmetry can be exploited to simplify the FEA model for appropriate cases. You must be careful that all components of the model (geometry, constraints, loads, materials, ...) are truly cyclic. Spring and mass elements will be useful in some types of models. They are easy to apply, although spring stiffness properties and orientation may take some thought. Modal analysis to determine natural frequencies and mode shapes is also quite simple. Remember that you don't need any applied loads for a modal analysis. In the analysis of assemblies, contact regions can be defined to handle the kinematic constraints between contacting surfaces.

There are some challenging questions and exercises at the end of this lesson that will require some exploration of these topics.

In the next lesson we will have a brief look at something totally different - thermal modeling in MECHANICA. This will include steady state and transient models, and calculating a thermal load to bring back into MECHANICA Structure to compute thermally induced stresses.

Questions for Review

1. When can you use cyclic symmetry?
2. Find out if there are restrictions on the orientation of the cyclic symmetry axis.
3. What are the two kinds of springs?
4. What properties are required to define a spring connection between two different points on the model?
5. What properties are required to define a spring connection between a point on the model and the ground? Where/how is the ground connection located?
6. To which of the following can cyclic symmetry constraints be applied: points, edges, surfaces.
7. Is it necessary to apply a radial constraint in a model with cyclic symmetry?
8. Is it possible to define a spring whose stiffness varies with its extension? What is this kind of spring called?
9. In general, what happens to the spacing between natural frequencies as they increase?
10. What is the **Load Interval** setting used for in a contact analysis?
11. How could you determine the size of a gap formed between two contact surfaces?
12. What happens if you accidentally specify two surfaces as a contact pair that never actually touch each other?
13. In the IPS system, what are the units of mass? Inertia?
14. How would you interpret the torsional stiffness of a spring connecting two points?
15. What happens to mating surfaces in an assembly if they are not designated as contact surfaces?
16. What happens if you try to form a pair of contact surfaces from concentric cylinders of different diameter?
17. Does a contact pair of planar surfaces allow friction between the surfaces?

Exercises

1.   Find out if it is possible, in an assembly, to create a spring of zero length by selecting points (on different components) that happen to be coincident.

2.   Create a model consisting of a point mass (1 kg) supported by a single spring (100 N/mm). What is the natural frequency of this simple system? What is the theoretical value? Set up a sensitivity study that will allow you to create a graph of natural frequency as a function of spring stiffness. (HINT: create a Pro/E parameter for the stiffness value.)

3.   Can you devise a method to create an assembly which contains a pre-loaded spring?

4.   Call up our 3D centrifuge model created using shells at the end of Lesson 7. Modify this to use cyclic symmetry and compare results with the previous model.

5.   Load our 3D frame model (**frame3d**) from Lesson 8 and find the first four modes of vibration (frequency and mode shape). Plot the mode shapes.

6.   Find out what happens if you try to define a contact surface pair between two surfaces that are in interference. For example, if the pin is slightly larger (by a few 1000^{ths} of an inch) than the hole in our assembly model.

7.   Find the first 4 natural frequencies of a thin (1 mm) circular plate with a diameter of 300 mm. The plate is made of aluminum. How do these frequencies depend on the edge condition: clamped (no rotation) versus free (rotation allowed)? What type of elements must you use to answer this question?

8.   For the two beams model, find the effect of the connecting spring stiffness on the load carried by the upper beam. For the most extreme values of this stiffness (zero and infinite), what load should be carried by the upper beam?

9.   Create a model of a musician's tuning fork. Find a real one and using its dimensions, see how accurately you can determine its frequency of vibration. Is the fundamental mode the desired mode for "tuning"? A common tuning fork used by guitarists produces a tone of the note A above middle C (440 Hz.).

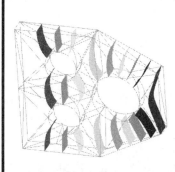

Chapter 10 :

Thermal Models

Steady state and transient models; transferring thermal
results for stress analysis

Synopsis

Overview of MECHANICA Thermal; discussion of boundary conditions, heat loads, units; solid
and plate models (steady state); transient models with time varying heat loads; transferring
thermal loads to Structure to compute thermally induced stresses

Overview of this Lesson

This lesson will give you a very brief introduction to the main features involved in running
MECHANICA Thermal. We will first have a quick overview of the program and discuss the
different model types available. Then we will look at the various boundary conditions and heat
loads that can be applied to the model. One of the most confusing aspects of using Thermal is the
topic of units - some material is presented to help organize this important topic.

After these preliminaries are over with, we will look at four examples: a solid model, a 2D plate
idealization, a transient (time dependent) problem, and finally the computation of a temperature
field in a solid model to be passed back to Structure as a thermal load. Most if the underlying
procedures for setting up these models, design studies, and result windows should be familiar to
you (if not, return to Lesson #3), so they will not be presented in fine detail. There are some new
result window types available for thermal problems, and some opportunities for making further
use of exported result data.

Overview of THERMAL

In this section we will present a fairly terse "birds eye" view of MECHANICA Thermal,
including what problems it can solve, the types of models available, and the variety of boundary
conditions and heat loads that can be applied to the model.

There are two main challenges in learning how to use Thermal effectively. First, you may find
that, unlike stress analysis using Structure, your physical intuition is challenged a bit more in
anticipating or interpreting the results. For a lot of users, the realm of heat transfer is not usually

familiar territory. Therefore, you are encouraged to spend additional time when going through the examples below to explore what happens to the model when you make alterations and to try to reconcile those results with your physical understanding of the problem. By all means, make up your own problems based on your own experience.

The second challenge involves the variety of different physical attributes that are either required or reported. In stress analysis we are interested in load and displacement (stress and strain), and really only have two physical material parameters (Young's modulus and Poisson's ratio) to worry about. In thermal analysis, although the primary solution variable is temperature, we are often more interested in heat flux through the model and, in particular, heat transfer through the surface. There are numerous parameters that affect these results: material properties (thermal conductivity, density, and heat capacity), surface conditions (specified temperature or insulated), applied surface heat loads, convection heat transfer at the surface, and so on. Keeping track of these factors, and in particular their units, will be a challenge. We will spend some time discussing units a bit later in this introduction.

What can THERMAL do?

MECHANICA THERMAL is a program that solves the equations involving heat conduction in solids. These solutions can represent a steady state condition, or can be time dependent (transient). The primary solution variable is the temperature of the solid. The solution incorporates or accounts for the physical attributes and processes shown in Figure 1 below. Thermal will solve for the internal and surface temperature wherever it is not explicitly prescribed in a boundary condition.

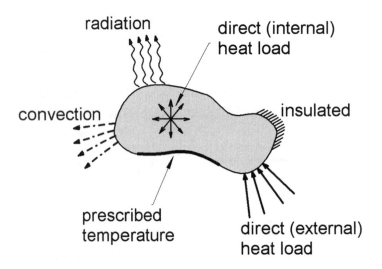

Figure 1 Basic physical attributes and processes in a THERMAL model

As indicated in Figure 1, the model can be subjected to numerous conditions on its interior and surface. These are

- **Direct heat addition/removal**. Heat energy can be directly added (or removed from) the model throughout its interior (think of heat generation in a nuclear fuel rod, or perhaps by microwave heating). Heat can be transferred through its surface by specifying a heat flux which is either uniform or can vary across the surface.
- The surface could be subjected to a **prescribed temperature**. This can be either uniform or vary across the surface.
- In **convective heat transfer**, the temperature of the surface is not known *a priori*, but we specify a convective heat transfer coefficient and the temperature of the surroundings. The surface temperature is then obtained as part of the solution.
- In **radiative heat transfer**, the surface temperature is obtained as part of the solution. Note that radiative heat transfer is a highly non-linear phenomenon. Although it is not available directly in integrated mode (as it is in independent mode), a work-around is possible using an iterative procedure and a "tuned" convection heat transfer coefficient.
- **Any model surface not specifically subjected to any of the above is assumed to be insulated.** This means there is no heat flux through the surface (but along the surface is OK), and the surface temperature is computed as part of the solution.

Why use THERMAL?

Based on the computed temperature field, Thermal can compute the heat fluxes within the model and through the surface(s). Obtaining these heat fluxes is often the primary motivation for executing the model. For example, it may be desired to design a cooling system for an electronic component. The design goals would be not to exceed a maximum allowed temperature of the component and to obtain a required rate of heat rejection from the device into the surroundings. These goals would be influenced by the geometry, material, and surroundings of the system. Sometimes the temperature distribution itself is of interest, as when it is desired to determine the best location for mounting temperature sensors on a system, and to estimate the relation between the measured temperature at the sensor and the temperature at another critical but inaccessible point. Finally, the temperature of the model can be returned to MECHANICA Structure as a thermal load which, via thermal expansion, can induce (sometimes very high) stresses in a part.

Material Properties

The temperature distribution within the solid is governed by Fourier's law of conduction which relates the heat flux through the solid to the gradient of the temperature. Thus, within the solid, the primary material property of interest is the *thermal conductivity*, the key parameter in Fourier's law. All materials in the MECHANICA materials library have values for this parameter. The default materials are isotropic, although custom materials can be created that have different conductivity in different directions. For transient problems, the ability of the material to absorb or hold thermal energy is governed by another material property, the *specific heat capacity*. This is also stored in the materials library (along with the *mass density*). Finally, when a temperature field is passed back to Structure to compute thermally induced stresses, the solution will depend on the value of the *coefficient of thermal expansion* of the material. Thus, we see that there are four material properties of interest, rather than the two required for Structure.

Model Types and Idealizations

As in Structure, the default model type in integrated mode is the **3D solid**. This is by far the most straight forward model to deal with (see the discussion of Units below!). However, as we have seen, quite often the essence of a problem can be represented by a simplified model type that captures the central features of the problem and can produce results much more efficiently. There are three additional model types available in Thermal:

2D Plate
- thin flat plate in XY plane
- model with shells created on surfaces
- analogous to plane stress models in Structure

2D Unit Depth
- conduction perpendicular to model negligible
- model with shells and solids
- analogous to plane strain models in Structure

2D Axisymmetric
- on X>0 half of XY plane; symmetry about Y-axis
- model with shells and solids

In this lesson, we will only look at 3D solid and 2D Plate models.

Of course, symmetry can play an important role in the solution of thermal problems, as it did in Structure problems. Most often, unless the temperature on a symmetry plane is to be specified, the appropriate symmetry boundary condition is obtained by doing nothing. Since there can be no heat flux through the symmetry plane, it must be treated as insulated. Cyclic symmetry is also available in Thermal models.

As indicated above, model idealizations are also possible in Thermal using beams and shells. **Beams** are strictly a 3D idealization shortcut. **Shells** can be used in both 2D and 3D problems. Treatment of these idealizations is quite far beyond the scope of this lesson. There is considerable on-line help available if you need to investigate these further.

More on Boundary Conditions

There are four possible surface boundary conditions that more-or-less directly involve the surface temperature. These are:

Specifed Temperature

The surface temperature is specified. It can be uniform or vary across the surface.

Convection

This condition requires the specification of a convective heat transfer coefficient and an external "bulk" temperature. The convection heat transfer can result in heat fluxes that either add to or remove energy from the part. The heat flux is related to the surface temperature according to

$$Q = hA(T - T_B)$$

where (in SI units)

Q = heat transfer (W = J/s = N.m/s)
h = convection coefficient (W/m^2.$^\circ$C)
A = surface area (m^2)
T = surface temperature ($^\circ$C)
T_B = bulk temperature ($^\circ$C)

Radiation

Radiative heat transfer is available in independent mode only. However, it can be simulated using an iterative procedure (since it is non-linear) and a "tuned" value of a convective heat transfer coefficient. It is governed by the following equation

$$Q = \varepsilon \sigma A (T^4 - T_\infty^4)$$

where (in SI units)

Q = heat transfer (W = J/s = N.m/s)
ε = emissivity
σ = Stefan-Boltzman constant (W/m^2.K^4)
A = surface area (m^2)
T = surface temperature ($^\circ$K)
T_∞ = enclosure temperature ($^\circ$K)

Insulated

Any surface entity that does not have one of the above three conditions applied to it is assumed to be insulated. This provides a boundary condition on the temperature gradient normal to the boundary - it must be zero. Thus, by Fourier's law, there is no heat flux normal to the boundary. The temperature of the boundary is therefore not specified beforehand, but is determined as part of the solution.

More on Heat Loads

Heat energy can be added directly to the model in several ways. These can be either internal or surface loads. Heat loads can be applied to volumes, surfaces, or edges. Depending on the model type and entity selected, the interpreted units of the heat load will be different.

A Note about Units

As indicated earlier, the interpretation and proper use of units is one of the more challenging aspects of using Thermal. Of course, if the units aren't used correctly, the results will be meaningless. You must be very clear about units for
- material properties
- convection coefficients
- heat flux
- heat generation

The units for these will change between model types and even entities within models.

IMPORTANT

It is strongly recommended that you avoid the use of British gravitational units (IPS) or similar. In thermal problems, this is in invitation to disaster. This system invokes yet another derived quantity (the Btu, aka British Thermal Unit) that serves only to confuse the issue (not to mention the confusion between weight and mass in that system). Try to work only in SI or mmNs unit systems. It is unfortunate that the Pro/E and MECHANICA defaults (unless reset) continue to use the old system of units.

The following units apply to 3D models. In particular, notice the difference in meaning between energy, heat transfer (energy per unit time), and heat flux (heat transfer per unit area). The on-line documentation is sometimes pretty loose in using these terms. For example, at various places, the on-line documentation uses the symbol "Q" to represent heat flux.

Quantity	Units (mm - N - s)	Units (m - kg - s)
length	mm	m
mass	tonne	kg
density	$tonne/mm^3$	kg/m^3
force	$tonne.mm/s^2 = N$	$kg.m/s^2 = N$
energy	$N.mm = mJ$	$N.m = J$
power, heat transfer (Q)	$N.mm/s = mJ/s = mW$	$N.m/s = W$
heat flux (q)	mW/mm^2	W/m^2
thermal conductivity (k)	$mW/mm.°C$	$W/m.°C$
convection coefficient	$mW/mm^2.°C$	$W/m^2.°C$

If you solve only 3D solid models, then you can use the units in the table above and you probably won't have much trouble. For example, tabulated values of convection coefficients, conductivities, and so on, are reported for the 3D world and correspond to the units shown in the table.

The difficulty in Thermal comes when we set up alternate model types or apply loads in different ways. For example, when we apply a convective heat transfer coefficient to the edge of a 2D Plate model, in order to account for the thickness of the plate model, the coefficient must be expressed in units of (in mmNs system) mW/mm.°C. Similarly, an interior heat load in a 3D model applied to a beam element must be expressed in terms of power per unit length of the beam (mW/m). To create these alternate unit versions of coefficients and loads, we must alter their numeric value. If this is not done correctly, of course our results will be wrong. We will explore this aspect of Thermal in the first two models.

For more information on the use of units, see the extensive on-line help.

Well, that's probably more than enough preliminary discussion. Let's get started on our first model.

Steady State Models

3D Solid Model

Launch Pro/E and create the model shown in Figures 2 and 3. Make sure you use the *mmNs* system of units. Sketch the section on FRONT and extrude backwards so that the front face is on the XY plane of the default coordinate system (we will need this geometry for the plate model later). The plate is 20 mm thick.

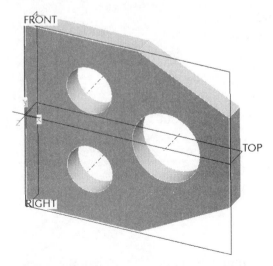

Figure 2 Solid model (note location of default CSYS)

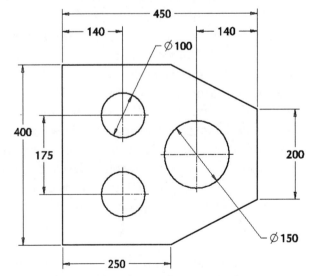

Figure 3 Section shape for model #1 (units mm)

Save the model and transfer into MECHANICA Thermal using

> ### *Applications > Mechanica*

In the **Model Type** window, change the **Mode** pull-down list to *Thermal*. Open the *Advanced* button to see the other options there. Accept the default (3D) by selecting *OK*.

You will notice a few new icons appear in the right toolbar, replacing some of the tools used in Structure. These are shown in Figure 4 at the right. Some of these buttons have fly-outs to allow you to create model entities on different types of geometry (surfaces, edges, points). Each of these toolbar buttons has equivalent commands available in the pull-down

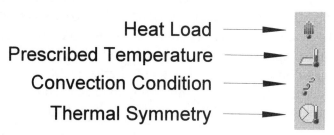

Heat Load

Prescribed Temperature

Convection Condition

Thermal Symmetry

Figure 4 THERMAL toolbar buttons

menu under *Insert* (in which case, the fly-out is handled by a small selection menu for points, edges, and surface options). The *Thermal Symmetry* tool allows only cyclic symmetry. Other tools in the toolbar are the same (*Define Materials*, *Measures*, etc.).

Select the *Define Materials* toolbar button. Pick the material copper (CU) and move it to the model list. Select *Edit* and check out the properties defined under the **Thermal** tab. Close the material editor, and select *Assign > Part*. Click on the part, then middle click twice to return to the graphics display.

You can turn off all the datums since they will not be needed. Note the orientation of the WCS.

Select the *Prescribed Temperature* tool. The boundary conditions in Thermal are also organized in sets. The first one is BndryCondSet1. Enter a name like *[t500]* for the prescribed temperature. The **Surfaces** button is already selected. On the model, pick the surface on the left face of the plate. Middle click. In the **Value** field, enter a value *500* (°C). Open the *Advanced* command and check out the other available Spatial Variations. There are options here to let you specify a prescribed temperature that is a function of position, or interpolated over the chosen entity. Leave this set to *Uniform*. See Figure 5. Accept the dialog window and a prescribed temperature icon will appear on the model.

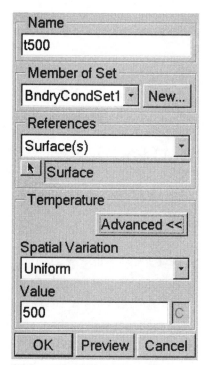

Figure 5 Creating a prescribed temperature boundary condition

At the narrow end of the plate, we will place a convection boundary condition. Select the *Surface Convection Condition* (this may be on a fly-out) tool in the right toolbar (also available in the pull-down menu). Enter a name for the condition; note that it is still in the same set as the prescribed temperature. Pick the right end surface of the part. Leave the spatial variation set to

Uniform, and enter a value for the convection coefficient h = *5*. Although not indicated in the window, the units of this are (mW/mm^2.°C). Leave the bulk temperature set to zero - this is the temperature of the medium receiving the heat energy from this surface (see the equation above). Notice that we could set a time varying heat transfer coefficient if desired. The completed dialog is shown in Figure 6. Accept the dialog. A new convection icon will appear on the right face of the model.

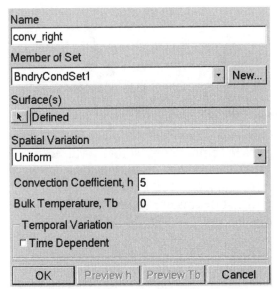

Figure 6 Creating a convection condition on a surface

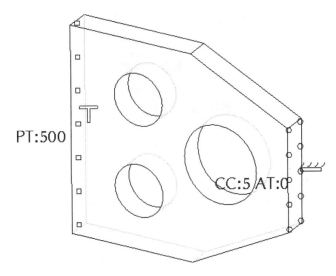

Figure 7 Final 3D plate model

All the other surfaces of the part that we have not set conditions on are assumed to be insulated. The completed model is shown in Figure 7.

Now, open the *Analyses/Studies* dialog window with

Analysis > Mechanica Analyses/Studies

Select *File > New Steady State Thermal*. Enter a study name *[plate3D]*. The constraint set is already selected. As usual, set up a *QuickCheck* analysis. Change the Output plotting grid to 10 (this will give us smoother contours). Verify your *Run Settings* and *Start* the run. Look in the *Study Status* listing for errors and/or warnings. AutoGEM will create something like 80 elements. Near the bottom, notice the new measures that are reported (heat flux, temperature gradient, max and min model temperatures). Assuming no errors, *Edit* the study to run a Multi-Pass Adaptive run with a convergence of 5% and a maximum edge order of 9. The MPA run will converge on pass 4 with a maximum edge order of 5. The minimum temperature is 21.6 degrees.

Now to create some result windows. Open the *Review Results* window. Insert a window definition to create a fringe plot of the temperature - this is the default. Under the *Display Options* tab, select *Show Edges*. Select *OK and Show*. See Figure 8. Notice that the fringe legend pallete is a bit different from the normal fringe legend. Select *Info > Model Min* to locate and label the minimum model temperature on the right end.

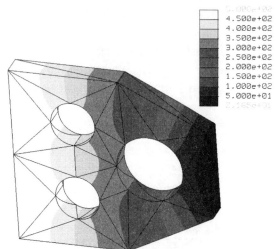

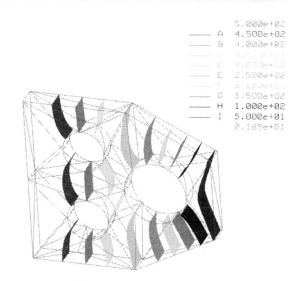

Figure 8 Temperature fringe plot

Figure 9 Temperature isotherm surfaces

Edit the temperature fringe plot. Go to the **Display Options** tab and select *Contour*, and *Isosurfaces*. Show this window (Figure 9). Observe that the isotherms are basically straight through the plate, as anticipated (all cross sections through the thickness should be identical.

Now create a new window called *[flux]*.
Select the following:

> *Display Type (Vectors)*
> *Quantity (Flux)*
> *Component (Magnitude)*

In the **Display Options** tab, select the *Animate* option. Show the window (Figure 10). What you see now is an animated representation of the heat flux through the part. Notice that the flux vectors are scaled and colored according to the magnitude of the local heat flux. Also, the vector direction is perpendicular to the isotherms (a consequence of Fourier's law).

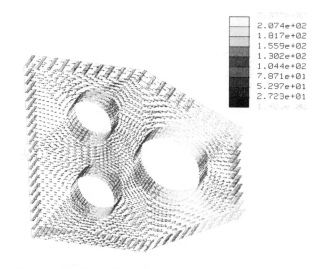

Figure 10 Heat flux vectors

Go to the FRONT view of the flux vector window. If you zoom in on any of the insulated edges, you should expect to see the flux vectors running tangent to the surface. Any indication of a flux vector not tangent to the surface represents a numerical or modeling error. We will examine this in more detail in the next 2D model.

We have now seen the main features of steady state thermal modeling for a 3D solid model - boundary conditions, convective loads, and prescribed temperatures, and some new variations available in the result window. We will now have a quick look at a different model type, and some additional result review tools.

2D Plate Model

Close all the result windows and return to the main display. In the pull-down menu, select

Edit > Mechanica Model Type > Advanced

Select the **2D Plane Stress (Thin Plate)** option. Select the button under **Surfaces**, then pick on the front surface of the part. Middle click. Now select the button under **Coordinate System** and pick on the default coordinates system. Accept the dialog. Because we are changing model type, we must confirm the change (since a number of modeling entities are lost). The surface is now highlighted in purple.

We must recreate the boundary conditions on the prescribed temperature and convection ends. Before we do that, we will create a shell on the chosen surface. This is identical to the creation of a model for plane stress analysis. Select the *New Shell* button, then click on the front surface of the part. In the **Shell Definition** window, set a thickness of 20 (mm). The material may already be defined, otherwise bring in the material copper. Accept the dialog. The shell idealization will be listed in the model tree.

Select the *Prescribed Temperature* toolbar button. Call it *[t500]* as before. Set the **References** type to *Edges/Curves*, select the button and pick on the left edge of the shell. Enter the value *500* and accept the dialog.

The creation of the convection condition is essentially the same, with one unexpected twist due to the way Thermal handles convection coefficients on edges. Select the *Edge/Curve Convection Condition* button in the right toolbar. Enter a name, like *[rightedge]*. Select the **Curves** button and pick on the right edge of the shell. Leave the spatial variation set to *Uniform*. For the value of the convection coefficient, we must be very careful.

Recall that for the previous 3D model, our value of the convection coefficient was 5 $mW/mm^2.°C$. This represents heat flux (mW) per degree per unit area of the surface. In the 2D plate model, we are specifying a coefficient along a linear entity only, and our units will be heat flux per degree per unit length along the edge. We must therefore account for the thickness of the plate - each millimeter along the edge corresponds to 20 mm^2 of actual plate surface. Thus, we must multiply the 3D heat flux value by the plate thickness. So, in the dialog window, enter a value of 100 for the convection coefficient. See Figure 11. It is unfortunate that Thermal cannot either show us the units being used, or can do this calculation on its own (since it already knows

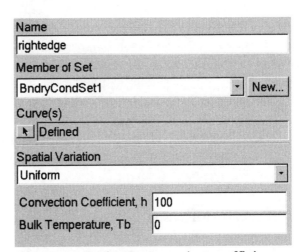

Figure 11 Setting a convection coefficient on a curve in a 2D plate model

the shell thickness!). In fact, if you leave the cursor in the data entry field for h, a pop up window indicates the expected units are heat per unit time per degree per unit area (which is wrong!). This is an example of the difficulties that can be caused by units in the Thermal module. Leave the bulk temperature set to 0.

Our model is now complete.

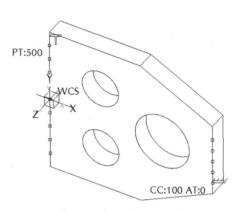

Go ahead and create a new steady state thermal design study. Set up a *QuickCheck* analysis with a plotting grid of 10. Make sure that in the *Run Settings* you select **Create Elements During Run** (otherwise, Thermal will try to use the 3D elements of the previous run and give you a fatal error). Run the analysis and look for messages in the *Study Status* list. AutoGEM creates just 18 plate elements. If all looks satisfactory, *Edit* the analysis and change it to an MPA with 5% convergence, max order 9. Run the study.

Figure 12 Completed 2D Plate model on front surface of part

The MPA converges on pass 4 with a maximum edge order of 4. The minimum temperature reported is 20.9°C - within a degree of the solid model solution.

Create some result windows to show the temperature fringe plot and the flux vectors.

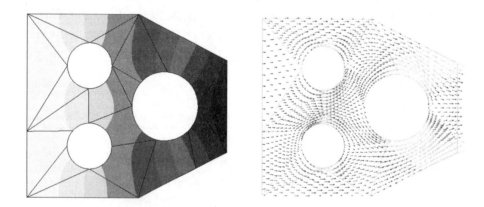

Figure 13 Results from 2D Plate model - temperature fringe (left) and flux magnitude vectors (right)

We should have a look at some data to study the behavior of the solution. Create a couple of convergence plots - for example for the maximum flux magnitude and the minimum model temperature.

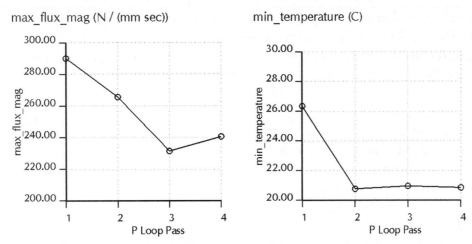

Figure 14 Convergence of flux magnitude and minimum temperature in 2D Plate model

Let's have a detailed look at the temperature along the right edge. Set up another graph as follows:

> ***Display Type (Graph)***
> ***Quantity (Temperature)***
> ***Graph Location (Curve)***

Select the button under ***Curve***. The screen shows an image of the 2D shell (Figure 15). Click on the right vertical edge. Middle click and note the message that indicates how the graph will be set up. Accept this, then show the result window. See Figure 16. The temperature minimum is at the center of the plate, and is essentially symmetrical as expected. According to this temperature distribution, what direction should the flux component in the Y direction be going along this edge? Create a graph window that shows the Y component of the flux along this edge.

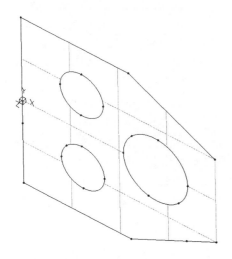

Figure 15 Selecting a curve for the temperature graph

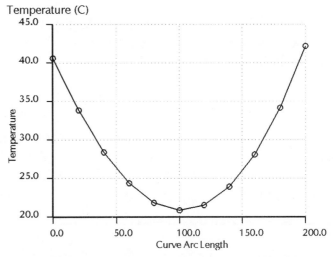

Figure 16 Temperature along right vertical edge

We will now set up another graph to perform an auxiliary calculation on this solution data. We would like to find the value of the net heat transfer coming out the right end of the model - this is the heat transferred to the surroundings. We can obtain that from the heat flux leaving the right edge of the part. We are interested only in the X component of the flux, since the Y component runs parallel to the surface.

Copy the temperature graph result window definition to another graph called *[Xflux]*. Change the **Quantity** to *Flux*, and **Component** to *X*. Select the same curve we had previously and show the graph. This graph is also reasonably symmetrical as expected. In the graph display window, select

       File > Export

This gives a number of different formats for the exported data. If you export it in spreadsheet form, you can easily modify the spreadsheet to integrate (perhaps by using trapezoidal integration) the flux with respect to the length along the edge. This value represents the total heat transfer per unit thickness of the plate. Once again you must be very careful with units. The flux data is given in units of (N/mm.s = mW/mm^2). The integral of flux with respect to edge length then has units of (mW/mm). Multiplying by the thickness of the plate finally gives us the heat transfer out of the plate in milliwatts. The value should be approximately 500 Watts.

Another interesting graph to plot is the Y component of heat flux coming out the top horizontal edge of the model. Since this edge is insulated, there should be no heat flux through the surface (ie in the Y direction). Create this graph. See Figure 17. Observe that the flux is not, in fact, exactly zero which represents a small modeling error. It is considerably smaller than the X component, however.

You might like to find out how to plot the flux in a radial direction through the edges of the circular holes. This should also be zero.

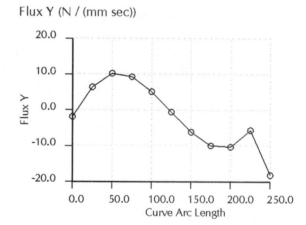

Figure 17 Y component of flux through the top edge (should be zero)

In the above, we have used all AutoGEM defaults. It is left as an exercise to revisit some of these models to see what the effects of changing the AutoGEM limits, or adding some mesh controls, would be. As always, we are interested in finding out how the model responds to various settings and run options.

Transient Analysis

A transient, or time dependent, analysis will allow you to find out how quickly (or slowly) the model reacts to the given boundary conditions. If the boundary conditions are constant, the transient analysis will eventually arrive at the steady state solution (as we calculated in the previous models). If the boundary conditions are also time dependent, then a transient analysis lets you observe the complicated time-dependent response of your model.

There are some restrictions on what you can do with a transient analysis. First, only 3D solid models can be treated. No shell or beam elements are allowed. The material(s) must be isotropic.

We will do a simple example of a transient analysis. The model consists of a long thin steel rod (300 mm long, 30 mm diameter). This model is shown in Figure 18. Create that now, making sure your units are set to mmNs.

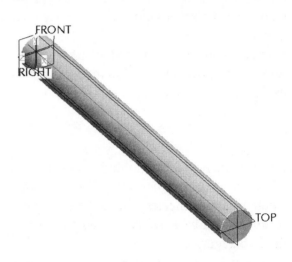

The rod is initially at a uniform temperature of 20°C. One end of the rod is maintained at the initial temperature. Starting at time t=0 seconds, the other end of the rod is exposed to a very hot (1000°C) environment, through a convection heat transfer coefficient of 0.5 mW/mm^2.°C. The remaining cylindrical surface of the part is assumed to be insulated. The primary questions we want answered concern the eventual temperature of the hot end of the rod, and the time it takes to reach that steady state.

Figure 18 Circular steel rod for transient analysis

Once the part is made, transfer into Mechanica with

> *Applications > Mechanica > Mode(Thermal) > OK*

Select the *Define Materials* tool and assign the material **STEEL** to the part.

Define a constant temperature surface at the left end using the *New Prescribed Temperature* tool. Call the condition *[tcold]* in BndryCondSet1. The default **Reference** is **Surfaces**. Pick on the left end of the rod, and enter a value of *20* (°C).

Define a *Surface Convection Condition* on the right end. Call it *[hotend]* in BndryCondSet1. Pick the surface at the right end. Enter a convection coefficient of *0.5* and a bulk temperature of *1000*. See Figure 19.

The model is now complete for a steady-state analysis. Open the *Analyses/Studies* window and select

File > New Steady State Thermal

Call the analysis *[trans_rod_SS]* (for steady state). Set up a *QuickCheck* analysis, with a plotting grid of *10*. Check the run settings, and *Start* the study. Open the *Study Status* window and look for errors. If all is satisfactory, *Edit* the study to run a MPA with a convergence of 1%, max order 9. Run the new study. The run converges on pass 3, max edge order 3. The maximum temperature in the model is reported as 782°.

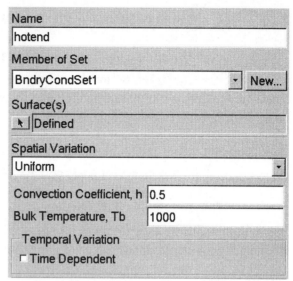

Figure 19 Defining the convection condition at the hot end of the rod

Create and show a result window for the temperature fringe. See Figure 20. This shows a series of equally spaced fringes. Recall that the default fringe legend uses equally spaced values. This fringe pattern indicates a linear temperature variation along the rod. This is confirmed by creating another result window that shows a graph of temperature along a curve extending the length of the rod (Figure 21).

Figure 20 Steady state temperature fringe plot

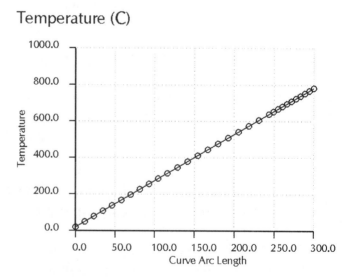

Figure 21 Variation of temperature along rod at steady state

Now for the transient analysis. Close down all the result windows and return to the
Analyses/Studies dialog. Select

> ### *File > New Transient Thermal*

Call this analysis *[trans_rod1]*. Enter the description (see Figure 22). Select the constraint set. In
the **Temperature** tab, set the temperature value to *20*. This is the initial uniform temperature of
the rod. Notice that the **Distribution** option allows you to specify a previously calculated
(MEC/T means MECHANICA Thermal) temperature distribution for the initial condition.

Open the **Output** tab. We will leave all defaults here. This is to use automatic time steps from
the start of the run (t = 0) until the steady state condition is reached. Accept the analysis
definition.

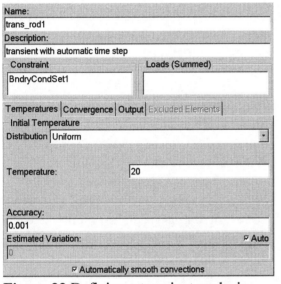

Figure 22 Defining a transient analysis

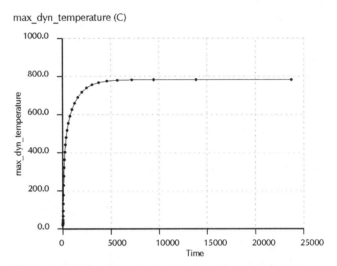

Figure 23 Maximum temperature in transient
solution

Start the transient analysis and open the *Study Status* window. The solution proceeds through a
number of (unequal size) time steps. If you scroll to the top of the window, you will observe the
edge order used for the solution, and the size of the first few time steps (quite small!). As the
solution proceeds, the time step is allowed to increase.

Create a result window. The default (and only) option is to show a graph. Select the button under
Measure, and select **max_dyn_temperature** in the list of defined measures. Show this graph -
see Figure 23. This shows that the solution changes very rapidly for the first few seconds, but by
5000 seconds has essentially reached its final value. If you double click on the data point at the
end of this graph, you will find that the reported temperature is 782°, and the final time is about
24,000 seconds. Leave the result windows.

Suppose you want to find the temperature distribution at a precise time. One way to do that is as
follows.

Create a new analysis called *[trans_rod2]*. Set the **Temperature** tab as before. In the **Output** tab, in the **Output Intervals** list, select *User-defined Output Intervals*. See Figure 24. Set the number of intervals to *40*. Then select *User Defined Steps*. Scroll down to the bottom of the list of points, enter a time value of *10000*, then press *Space Equally*. This will yield equal time steps of 250 seconds.

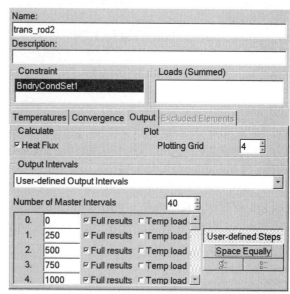

Figure 24 Creating a transient analysis with equal spaced intervals

Run the new analysis and open the **Study Status** window. You will observe that the number of actual steps is larger than we requested. However, on closer examination, notice that the run did stop at 10,000 seconds, with most of the time steps set to 250 seconds as requested. Early in the run, however, the time steps were considerably smaller in order to track the rapidly changing solution. The actual computation time steps are controlled by Thermal using a different setting in order to maintain accuracy. Very nice to know!

Create a result window to show a graph of the maximum dynamic temperature. See Figure 25. In the result window definition dialog, notice that you can select any particular time step and plot data for that particular instant in time. Observe that in Figure 24, the boxes "Full Results" were automatically checked for each time interval. Another interesting output result is obtained if you request an animated fringe of the temperature. Select *Contour* and *Isosurfaces*. When you show this, the isotherms will move down the rod, eventually reaching their equally-spaced steady state positions. The final configuration is shown in Figure 26.

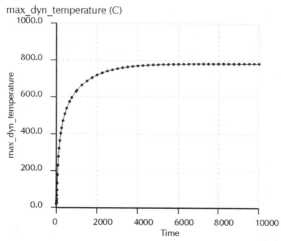

Figure 25 Maximum temperature in transient solution (equal spaced intervals)

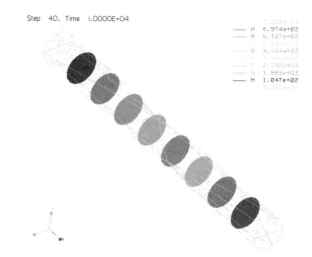

Figure 26 Isotherms at end of transient solution

You might also have a look at the flux magnitude in the rod at various time steps. Initially, the magnitude is essentially zero everywhere. At the final time step, the magnitude is essentially constant everywhere.

We will leave transient analysis now. One thing you might like to try on your own is to perform a similar transient analysis on the rod with a specified heat load on one end (replacing the convection) where the heat load is a function of time.

Thermally Induced Stresses

Our final example in this chapter is to show how a temperature field computed by Thermal can be passed back to Structure in order to calculate thermally induced stresses. The temperature field affects the Structural model through the effect of thermal expansion. If the model is physically constrained against this expansion (or contraction), very high stresses can be produced.

We will examine the simple model shown in Figure 27. Create this model called *[ring]* using the mmNs part template. The sketch for the shape is shown in Figure 28. Notice the four rounded corners (R50).

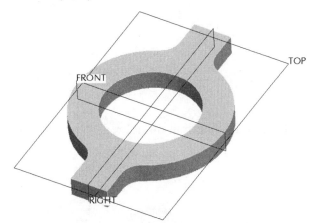

Figure 27 The ring model to explore thermally induced stresses

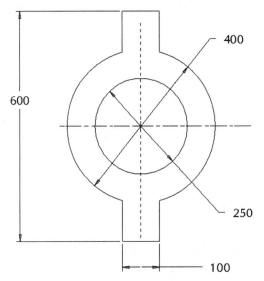

Figure 28 Section shape for ring model (units mm). Thickness = 50 mm

Creating the Thermal Model

When the model is created, transfer into MECHANICA Thermal using

Applications > Mechanica > Mode(Thermal) > OK

We will load the thermal model with two prescribed temperature surfaces and a convection condition on the inner cylinder.

First, select the *New Prescribed Temperature* tool. Name the condition *[hotend100]* and pick on the farthest front rectangular surface. Assign a temperature of *100* (°C).

Select the *Prescribed Temperature* tool again and create a condition *[coldend0]*. Pick on the farthest back rectangular surface and assign a temperature of *0*.

Select the *New Surface Convection Condition* tool. Name this condition *[inner_surf]*. Pick both halves of the inner cylindrical surface. Apply a convection coefficient of **0.25** (what are these units?) with a bulk temperature of *0*.

Finally, select *Define Materials* and assign the material CU (copper) to the part.

The model is now finished (Figure 29) and ready for analysis.

Create a *New Steady State Thermal* analysis called *[ring_temperature]*. Set up a QuickCheck analysis. This should create a mesh of about 70 elements.

IMPORTANT:

The same mesh must be used in both Thermal and Structure (in order to align the temperature data with the structural elements). For subsequent runs of this model, make sure the **Run Settings** option *Use Elements from an Existing Study* is checked. If you change the model geometry, however, you will have to check *Create Elements during Run* the first time you re-run any analysis in either Thermal or Structure.

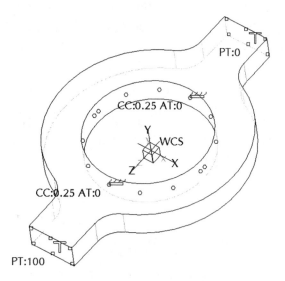

Figure 29 Thermal model ready for analysis

Assuming the QuickCheck succeeded, change to an MPA analysis with a convergence of 5%. Check the *Run Settings* and *Start* the analysis. The run should converge on pass 4 with a maximum edge order of 5. Nothing unusual here.

Create a temperature fringe plot. See Figure 30. The hot end shows as 100° with the cold end at 0° as desired. Notice that this model has two planes of symmetry. You might like to come back later and implement symmetry conditions (insulated surfaces) on a quarter model.

We are now ready to transfer to Structure where we will apply this temperature distribution as a thermal load.

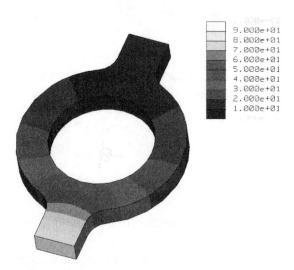

Figure 30 Temperature distribution on ring model

Creating the Structure Model

Select (in the pull-down menu)

Edit > Mechanica Model Type > Mode(Structure) > OK

First we will apply some displacement constraints. Select the *New Displacement Constraint* tool and create a surface constraint *[cold_end_fixed]* on the cold end. Fix all degrees of freedom.

Select the *Displacement Constraint* tool again and pick the hot end rectangular surface. We will constrain this against translation normal to the surface (Z direction), but must FREE the X and Y translations. We are going to set the reference temperature for the model at 0°C, so at the hot end the rectangular section will want to expand in both X and Y directions. If we constrained against this motion, we would find extremely high stresses induced here (try it later!) that would overshadow what is happening in the ring itself.

To apply the thermal load, we must start in the pull-down menu:

Insert > Temperature Load > MEC/T

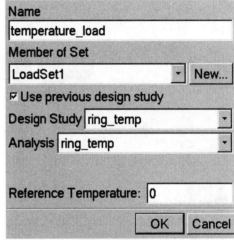

Figure 31 MEC/T dialog to assign a thermal load from a previous analysis

This opens a new dialog window. Name the load *[temperature_load]* in LoadSet1. The box beside *Previous Design Study* is checked. The study *ring_temp* that we did in Thermal has already been identified for us. The *Reference Temperature* at the bottom indicates the temperature at which there is zero stress induced in the model consistent with the current constraints. Leave that set to 0. The completed dialog window is shown in Figure 31.

The Structure model is now complete (Figure 32).

If you open the model tree, you can see all of the MECHANICA-related data listed there. Although you can get information about the thermal data at this time, you cannot edit or delete them.

Create a *New Static* analysis called *[ring_stress]*. Make sure the constraint and load sets are selected. Set up a **QuickCheck**. Make sure the *Run Settings* are correct (recall the note above about using the same mesh as the thermal analysis) and *Start* the run. Assuming there are no errors, change the analysis to an MPA with a 5% convergence and rerun the study.

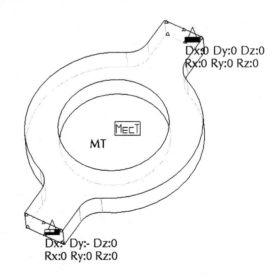

Figure 32 Structure model complete with thermal load

Open the ***Study Status*** window. The analysis converges on pass 6 with maximum edge order 6. The reported maximum Von Mises stress is 108 MPa. The maximum displacement in the X direction is about 0.2 mm. This is not much on something 600mm in diameter - it doesn't take much thermal load to produce fairly high stresses.

Create a couple of result windows that show the Von Mises stress fringe plot, and the deformed model. These are shown in Figures 33 and 34 below. Locate the position of the maximum stress in the model. What do you think would happen if we decreased the size of the rounds (or deleted them altogether)? Observe the expansion of the rectangular hot face. What would happen if we constrained this face against this expansion?

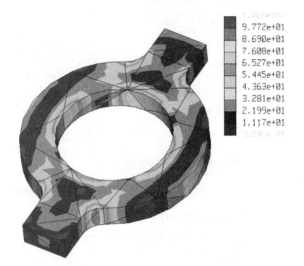

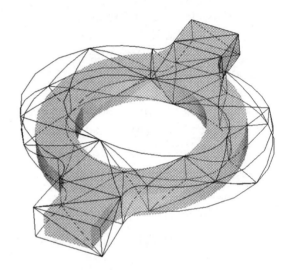

Figure 33 Von Mises stress due to thermal load

Figure 34 Wireframe mesh showing deformation due to thermal load. Note expansion of hot face.

Summary

This concludes our discussion of MECHANICA Thermal. This lesson has shown you the basic procedures and tools available for performing steady state and transient thermal analysis. We have explored only two of the four available model types (3D Solid and 2D Plate models), having missed the 2D Unit Depth and 2D Axisymmetric models. The unit depth model is analogous to the plane strain models in stress analysis, so many ideas should carry over from there. Axisymmetric models have similar restrictions as in stress analysis.

We looked mostly at steady state models here. Transient analysis can become considerably more complicated with the application of time dependent loads and boundary conditions. As usual, there are quite a few tools available that we have not discussed. Hopefully, having seen the general procedures presented here, you will be able to make sense of those tools as you explore them on your own.

As mentioned earlier in the lesson, developing a physical intuition for thermal analysis is a challenge since many people do not have a broad experience in this field. Furthermore, it is necessary to be very careful about units and data entry. It is quite easy to create program input without proper units and not be able to recognize this error in the program results. A lot of care and experience is necessary here.

Conclusion

This completes this tutorial, and we have covered a lot of ground. Even so, we have not looked at many MECHANICA commands, functions, or analysis types. It is hoped, however, that you are now comfortable enough using the program that you can experiment with these other capabilities on your own without getting lost and that you can more easily follow the on-line documentation. When you do try something new, you should set up a simple problem for which you already know the answer (either quantitatively or qualitatively) just to make sure your procedures are correct. Also, the on-line documentation is available to answer your questions about other aspects of MECHANICA. Your installation may also have a set of verification examples provided with the software.

When using Pro/E to create solid models, remember the discussion of Lesson 1. As we have seen many times here, the FEA model is not necessarily identical to the Pro/E CAD model, in fact it is usually not even close! You may be able to defeature the part[1], and certainly use symmetry whenever you can, in order to produce an efficient as well as effective FEA model. You should also be able to use idealizations to model some (or all) aspects of the problem at hand. These can greatly speed up the computation.

In closing, it is useful to remind ourselves of some of the comments made in the first two chapters. First,

"Don't confuse convenience with intelligence."

MECHANICA is undoubtedly a very powerful analysis tool but it will not automatically solve your problems for you. You should realize by now that, like all other computer tools, unless it is used properly, the results it produces can be suspect. Remember that in FEA we are finding an *"approximate solution to an idealized mathematical model of a simplified physical problem."* It is expecting a lot to hope for results that *exactly* match the solution found by nature! The most we should hope for are answers that are sufficiently accurate so as to be valuable.

---

[1] Defeaturing does not necessarily mean loss of data. Modern design practice, in fact, involves the use of FEA much earlier in the design process in order to do preliminary concept evaluation. In that case, it is likely that low level or finer detail has not been determined or included in the model yet.

Second, MECHANICA is a huge program that will take many, many hours to master. As your knowledge and experience grows, applying your new skills in increasingly more complicated problems is an inviting prospect. However, when you start to feel the urge to rush off to your computer to tackle a new problem, remember the first goal of FEA is to

> *"Use the simplest model possible that will yield sufficiently reliable results of interest at the lowest computational cost."*

It may even be that your problem can be solved in other (cheaper and quicker) ways.

In short,

> *"Let FEA become a tool that extends your design capability, not define it."*

Good luck, and happy computing!

Questions for Review

1. What is Fourier's Law? What material property does it involve? What are its units?
2. What is the (mathematical) consequence on the computed temperature distribution in a solid of specifying the following boundary conditions:
   (a) an insulated surface
   (b) a surface with a specified heat flux
   (c) a surface with convection
3. What four material properties are involved in a thermal analysis? Make a table of these and show their units in mmNs and SI.
4. What are the restrictions imposed on performing a transient analysis?
5. How do you ensure the same mesh is used for both Thermal and Structure? Does it matter which module creates the mesh?

Exercises

1. Explore the effects of various mesh control functions on the 3D Solid and 2D Plate models. In these models, describe the effect of mesh control on execution time, convergence, surface fluxes, max/min temperature, and so on.

2. Cut the 3D model in half along the horizontal symmetry plane. What boundary condition is required to implement symmetry? Do the results agree with the full model?

3. Find out if thermal problems are subject to the same sorts of singularity problems that occur in stress analysis.

4. How could you set up a model that would illustrate the effect of the material's specific heat capacity in a transient analysis? Try it! Does this property have any effect in a steady state analysis?

5. Create a model of a solid sphere. The sphere surface is subjected to convection. Perform a transient analysis from an initial uniform temperature state to a final state where the surface temperature is exactly half the initial value. How much of this solution can be verified by analytical means?

This page left blank.